An Introd

SECOND E

An Introduction to Soil Science

SECOND EDITION

E. A. FitzPatrick
Senior Lecturer in Soil Science
University of Aberdeen

Copublished in the United States with
John Wiley & Sons, Inc., New York

Longman Scientific & Technical
Longman Group UK Limited,
Longman House, Burnt Mill, Harlow,
Essex CM20 2JE, England
and Associated Companies throughout the world.

Copublished in the United States with
John Wiley & Sons, Inc., 605 Third Avenue, New York, NY 10158

First published 1974. This is a completely rewritten new edition of
AN INTRODUCTION TO SOIL SCIENCE, first published by
Oliver & Boyd 1974

Second edition 1986
Reprinted 1988

British Library Cataloguing in Publication Data
FitzPatrick, E. A.
An introduction to soil science. – 2nd ed.
1. Soils
I. Title
631.4 S591
ISBN 0-582-30128-9

Library of Congress Cataloging-in-Publication Data
FitzPatrick, Ewart Adsil.
An introduction to soil science.
Bibliography: p.
Includes index.
1. Soil science. I. Title.
S591.F54 1986 631.4 85-19790
ISBN 0-470-20670-5 (USA Only)

Contents

List of Figures

List of Colour Plates

List of Tables

Acknowledgements

I should like to thank those persons who have helped in the preparation of this book through their suggestions or by providing some material, especially Professors J. W. Parsons and J. Tinsley, Dr M. N. Court, the late Dr J. R. H. Coutts, Dr N. T. Livesey, Dr D. A. MacLeod and Dr C. E. Mullins: Mr I. G. Pirie for helping to prepare Figs 1, 2.1, 2.15, 2.17, 2.18, 3.10, 3.11 and to the following for permission to use various illustrations: The British Museum for Figs 2.2, 2.6; the Macaulay Institute for Soil Research for Figs 2.11, 2.12, 2.24, 3.3; Macmillan Publishing Co. Inc. for Fig. 2.21; Dr K. S. Killham for Fig. 2.25; Aerofilms Ltd. for Figs 2.34, 6.9, part of 8.7; USDA Soil Conservation Service for Figs 2.36, 4.4, 6.5, 6.10; the late F. M. Synge for Fig. 3.4; CSIRO, Australia for Fig. 3.6; Dairy Science Department, University of Florida for Fig. 6.4; Paul Popper Ltd for Fig. 6.11; Massey-Ferguson for Fig. 6.12; HMSO for Fig. 8.5; FAO for part of Fig. 8.8; John Bartholomew & Son Ltd. for the use of Bartholomew's Regional Projection in Figs. 8.10, 8.11, 8.12, 8.13, 8.14 and 8.15; Soil Survey of England and Wales for Fig. 9.1; Cambridge University for Fig. 9.2 and the Ordnance Survey for Fig. 9.3.

Introduction

After many centuries of the use and misuse of soils there is now a rapidly growing consciousness about the place of soils in the environment and their importance as vital factors in the life of most organisms. Hitherto soils have commanded little respect and seldom have they had optimum management. Perhaps it is true to say that soils are our major natural resource because most of our food and clothing comes directly or indirectly from them. Since most soils take thousands or even millions of years to form they cannot be replaced if they are washed away by erosion. It is therefore of paramount importance that our soil mantle be carefully nurtured so that it will be preserved in a healthy and fertile state for generation after generation. The archaeological evidence shows clearly the widespread destruction of soils by the many and diverse civilisations that have flourished and disappeared. Probably the downfall of many can be attributed to progressive erosion – the Mayan civilisation of Central America seems to fall into this category.

This book is intended for beginners who would like to know something about the nature and formation of soils and also about their uses and geographical variation over the surface of the Earth. Pupils in their final years at school should find this book useful, and it should also be helpful to the large number of biology, geography and geology students in their first year at universities and other institutes of higher learning. There are also a number of generally interested people for whom this should provide the breadth and depth of information they require.

Soils, or the pedosphere, are composed of air, water, mineral material, organic material and organisms, and can be regarded as an amalgam of the lithosphere, the biosphere, the hydrosphere and the atmosphere as shown diagrammatically in Fig. 1. The amount and type of these constituents vary widely from place to place over the Earth's surface causing an almost infinite variability

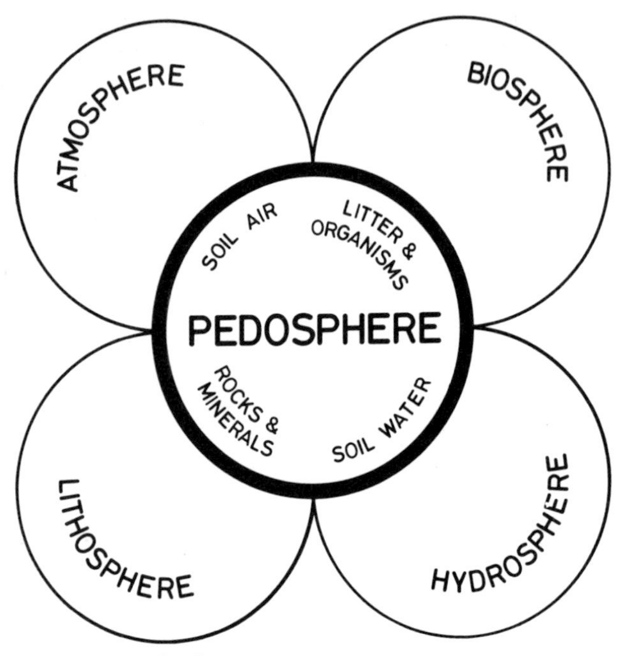

Fig. 1 The pedosphere

in the types of soils. Such a wide range induces many complications particularly with regard to classifying soils. On the other hand the wide variability in soils presents a challenge and holds an evergreen fascination for the soil scientist.

Soils can be studied in a variety of ways. They can be considered as natural phenomena and worthy of independent study or they can be studied in relation to the natural environment. Probably the greatest amount of study is directed towards establishing the distribution and mapping of soils, and determining their suitability for crop production, be they food crops or trees.

At the end of the text there is Appendix 1 giving the method for describing soils. In addition there is a glossary containing a number of the common terms used in soil science.

At present great emphasis is being placed on the utilisation of soils but before they can be utilised properly they must be recognised as highly organised physical, chemical and biological systems whose nature, properties, and place in the environment should be understood. From this fully integrated systems of soil, plant and animal management should be developed. These systems would range from the level of the farm which is an obvious example to the less obvious type such as tropical rainforests which can also be managed.

The FAO-UNESCO world soil map is being used for predictive purposes and for the economic feasibility of improving crop production particularly in underdeveloped countries.

1 *Fundamental concepts*

Most people usually think of soils as the upper few centimetres of the Earth's crust permeated by plant roots or cultivated. This is a somewhat limited approach which focuses attention mainly on soils as media for plant growth. In the first instance soils should be regarded as natural phenomena and part of the environment.

Soils as natural phenomena

The soil scientist recognises not only the top soil in which plants grow but also many other layers beneath, and the first step towards understanding soils is to dig a pit into the surface of the Earth and to carry out visual observations. The depth of the pit is determined by the nature of the soil itself and normally varies from one to three metres, below which is relatively unaltered material.

The pit reveals a characteristically layered pattern mainly expressed through differences in the colour of the individual layers. Each layer is known as a *horizon* and the set of layers in a single pit is termed a *soil profile*. The two examples given in Figs 1.1 and 1.2 illustrate the main principles concerned with soil profiles as well as some of the variability that exists between soils found under certain types of natural coniferous and broad-leaved deciduous forest.

Figure 1.1 and Colour Plate IVA show horizon sequences in Podzol profiles – soils of common occurrence in humid temperate areas under coniferous forest. At the surface there is an accumulation of freshly fallen plant litter, below which is dark brown, partially decomposed plant material, mainly inhabited by fungi, bacteria and small arthropods which are largely responsible for its break down. This grades into very dark brown or black amorphous organic matter in an advanced stage of decomposition. The progressive stages in the decomposition are visible in hand speci-

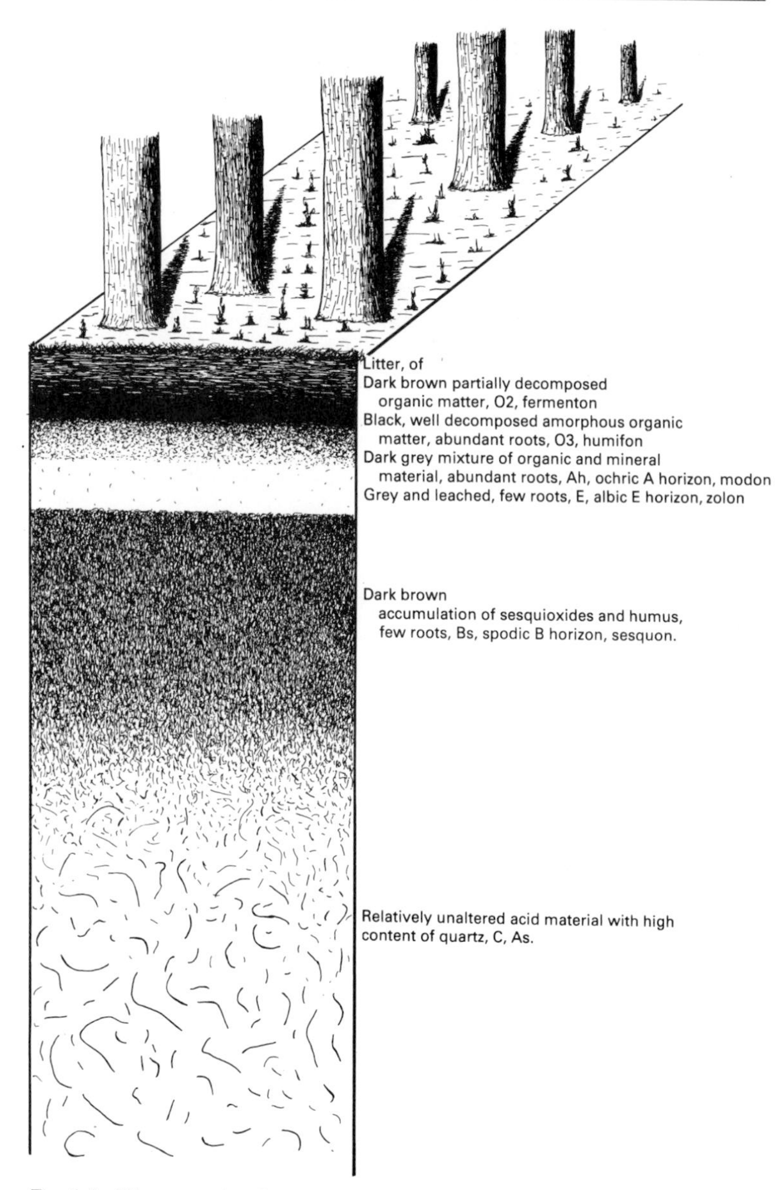

Fig. 1.1 Diagram of a Podzol profile which is the characteristic soil of the northern coniferous forests

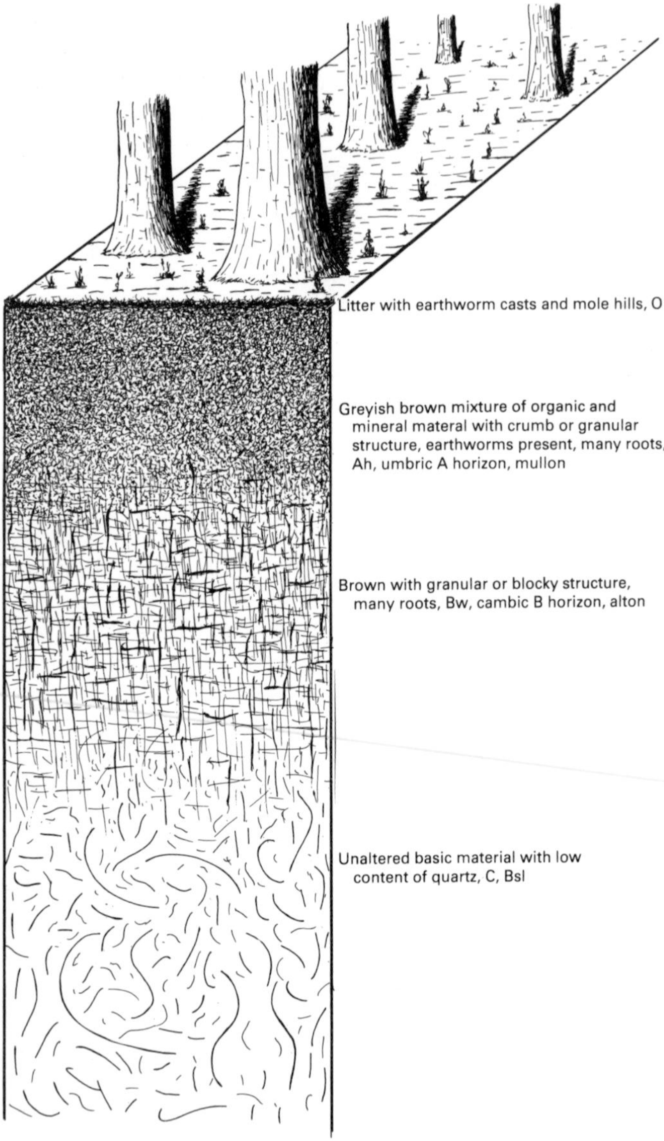

Fig. 1.2 Diagram of a Cambisol profile which is the characteristic soil of the temperate deciduous forests

mens but are more clearly seen in thin sections (see Figs 2.26 and 3.9). Beneath the organic layers there is a very dark grey mixture of black organic matter and light coloured bleached mineral grains, mainly quartz. This horizon is underlain by a pale grey or white horizon composed mainly of bleached quartz. The mineral material in these two horizons is strongly weathered and percolating water removes the colouring substances which are mainly compounds of iron. The substances removed from the upper horizons accumulate immediately below to form a middle horizon with its characteristic brown or dark brown colour. Finally there is the relatively unaltered material which is usually very sandy containing a high proportion of quartz and is therefore acid in composition.

Figure 1.2 and Colour Plate IB are examples of Cambisol (Inceptisol, Altosol) profiles that develop under deciduous forests and from basic material which is characterised by having minerals containing large amounts of basic cations. At the surface there is a thin loose litter of leaves and twigs containing numerous earthworm casts and mole hills. Beneath is a brown or greyish-brown horizon of organic and mineral material, intimately mixed mainly by earthworms. This horizon grades into a middle brown horizon showing less evidence of faunal activity, followed by relatively unaltered material. This soil has not the marked contrast between horizons as shown by Podzols because in this case iron is not translocated from the upper to the middle horizon.

The letters A, B and C are usually used to designate the upper, middle and lower horizons respectively, but these letters have different morphological and genetic connotations to different workers and their usage can at times be very misleading. However the system developed by FAO is given in Chapter 5 and included in the illustrations together with horizon names according to the USDA (1975) and FitzPatrick (1980).

Both Podzols and Cambisols are cultivated in a variety of ways leading to marked changes in the character of the upper horizons. Podzols are naturally infertile and need a considerable amount of treatment before agricultural crops can be grown successfully. On the other hand Cambisols have a high natural fertility because of the basic material and vigorous soil fauna and flora.

The soil profile is simply a two-dimensional section through soil which in reality extends laterally in all directions over the surface

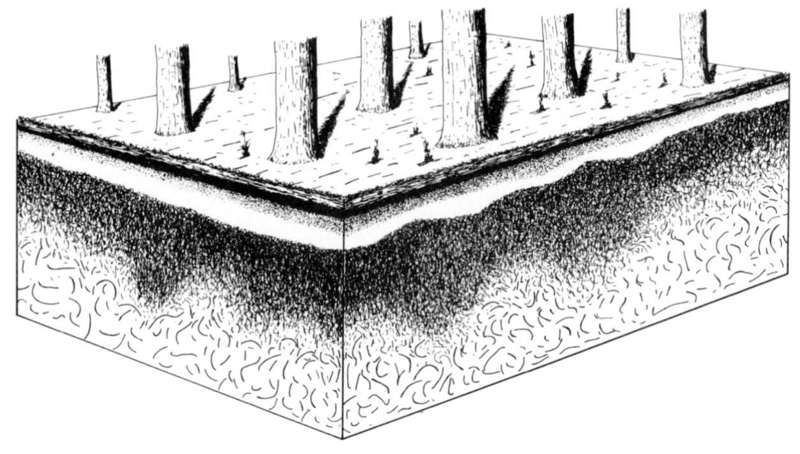

Fig. 1.3 A three-dimensional block of soil

of the Earth forming a three-dimensional continuum (Fig. 1.3). Further, the constituent horizons do not remain uniform throughout their lateral extent but exhibit a gradual change from one set of horizons to another.

A very important aspect of soils is that they seldom if ever have sharp boundaries, usually they grade gradually from one into another. This is often overlooked when isolated profiles in separate pits are examined. An example of the lateral changes in soils is given on page 191. Soils can be considered as the thin veneer that covers a considerable part of the Earth's surface and which is very vulnerable. The result is that on the majority of slopes there is some naturally produced stratification particularly in the older landscapes. Thus soils show both vertical and lateral differentiation of horizons.

Finally, it should be stated that soils are not static but are continually forming and changing so that they form a natural space-time continuum.

Soils as part of the environment

Soils can be considered as a product of the environment – the attitude taken by the pure soil scientist – or they can be regarded as a part of the environment, the more general attitude

taken by the natural scientist. When the latter attitude is adopted, soils are then part of a larger and highly complex system which for convenience can be broken down into a number of simpler cycles and relationships, the most important including;
The carbon cycle (page 68)
The moisture cycle (page 31)
The nitrogen cycle (page 69)
Energy relationships (page 28)
The oxygen cycle
The mineral cycle

Prior to the last century the influence of humans on these cycles could be regarded as relatively small but since the industrial revolution they have increasingly influenced their environment and in some cases exercise almost absolute control. Perhaps the classical example is through the increased growth of crops by the use of fertilizers and pesticides resulting in a general benefit to humanity, but the application of too many fertilizers, insecticides, or weed killers can lead to their presence in drainage waters which ultimately will pollute rivers and lakes thus negating the benefits from an increased food yield, because the water supply is rendered useless. Recently the enormous damage to crops, animals, property and humans by atmospheric pollution has received international attention. A few of these cycles will be discussed in Chapters 3 and 6 from the standpoint of minimum disturbance by people as well as mentioning some of the dramatic disruption that can take place when unforeseen changes in the soil are introduced. In the context of plant growth the term soil has been extended to mean the various mixtures that are used for growing plants in pots, greenhouses and the like.

In the following two chapters the factors and processes of soil formation are considered and together they illustrate that soils are formed by the interaction of certain specific environmental factors.

2 *Factors of soil formation*

Dokuchaev, the famous Russian soil scientist, showed that soils do not occur by chance but they usually form a pattern in the landscape, and furthermore he firmly established that they develop as a result of the interplay of five factors: parent material, climate, organisms, topography and time (Fig. 2.1). Below are presented the more important characteristics of each factor.

Parent material

Jenny (1941) defines parent material as "the initial state of the soil system". The precision of this definition cannot be ques-

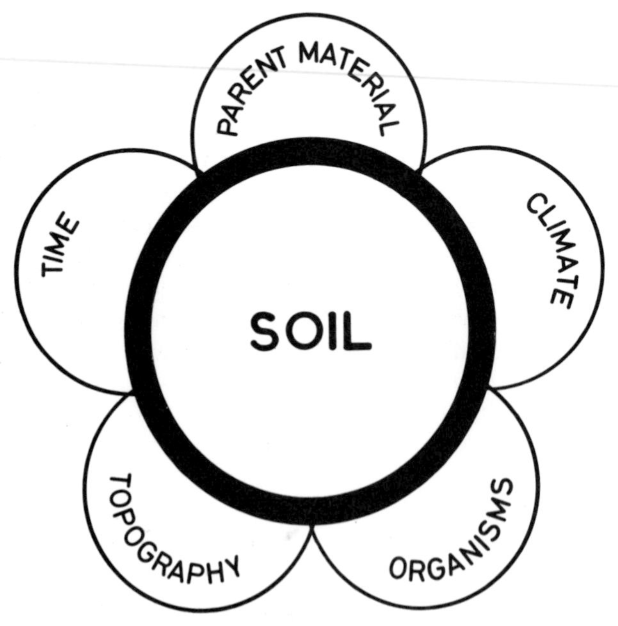

Fig. 2.1 Factors of soil formation

tioned but it is often difficult to determine the initial stages of soils, for in a number of cases the character of the original material has been changed markedly by soil formation.

Often the unaltered material in the lower part of the profile is similar to the material from which the horizons above have formed but this is not always the case. For example, there may be a thin cover of drift overlying rock. In fact, recent studies have shown that many, if not most soils display some evidence of layering hence the difficulties in applying the A, B, C designation of soil horizons.

Parent materials are composed of mineral or organic matter or both. The most widespread mineral materials range from consolidated rocks to various unconsolidated sediments. The consolidated rocks can be igneous, metamorphic or sedimentary.

The igneous rocks form by the cooling and solidification of molten magma and include granite and basalt. The sedimentary rocks are composed of weathering residues that have usually accumulated in large bodies of water. They range from almost pure quartz sand to limestone and chalk composed of shells or precipitates of calcium and magnesium carbonate. Many sedimentary rocks are cemented by silica, calcium carbonate or iron oxides.

Metamorphic rocks are those that have undergone secondary change and include both altered igneous rocks and sediments. The changes are brought about by the very high temperatures and pressures that occur at great depths in the Earth's interior. The range of metamorphic rocks derived from sediments varies from the weakly metamorphic rocks such as slates, through schists to the highly metamorphic gneisses. Some workers consider that many granites are an ultimate stage in metamorphism. Hornblende schist and gneiss are formed by the metamorphosis of basic igneous rocks.

There is only a weak correlation between the nature of rocks and the soils developed from them. The same type of rock can give rise to very different soils depending upon the nature of the other factors, particularly climate. Basalt may give a brilliant red, highly weathered soil (Ferralsol) in the humid tropics or it may give a black tropical soil (Vertisol) in a semi-arid environment. The best correlation is with texture. Highly quartzose materials tend to give very sandy soils while basic rocks and fine grained sediments tend to give fine textured soils. A better correlation is found in the early stages of soil formation when the general trend is for basic and calcareous rocks to give very fertile soils with less fertile soils on the acid rocks.

The chemical and mineralogical compositions of some common rocks are given in Table 2.1.

The unconsolidated material comprises a wide range of superficial deposits including the following:

Alluvium: a deposit of material carried in suspension by a river. It ranges from clay to very large boulders. Some of the most fertile land in the world consists of alluvial deposits.

Colluvium: crudely stratified material on slopes formed by soil creep or flow.

Dune sands: usually coarse sand formed by wind into ridges of various shapes. They are common along coasts and in arid areas (Fig. 2.34, see p. 51).

Glacial drift (moraine): the debris or material deposited by a melting glacier, it can vary widely in particle size and usually contains a mixture of rock types. It covers a large part of the northern hemisphere (Fig. 2.33, see p. 51).

Lacustrine clays: fine material sedimented at the bottom of lakes and now exposed at the surface due to a change in the nature of the drainage.

Loess: a deposit of fine material transported by wind, it is usually buff coloured, unstratified and forms vertical walls. Its source is variable, the loess in northern China has been derived from the Gobi desert while the extensive deposits in Europe and North America have been derived from the outwash material from melting glaciers (Fig. 2.35, see p. 53).

Marine clays: fine material sedimented at the bottom of the sea and now exposed at the surface due to change in sea-level. Very good examples occur in the fiords of western Norway where the land has risen up since the loss of the weight of glacier ice.

Pedi-sediments: material deposited by surface-wash on long, gently sloping plains. They consist of old soil material such as quartz gravel and concretions and are very common in tropical and subtropical countries.

Raised beaches: beach deposits that are now above the present high-water mark. They are very common in coastal situations

Table 2.1 Chemical composition of some common rocks

Rock type	*Composition (%)*											
	SiO_2	Al_2O_3	Fe_2O_3	FeO	MgO	CaO	Na_2O	K_2O	TiO_2	P_2O_5	MnO	H_2O
Basalt Labradorite, augite and pseudomorphs of chlorite and goethite after olivine; and opaque minerals.	53	18	5	2	4	7	4	2	1	<1	<1	2
Diorite Hornblende, biotite, titanite, oligoclase, quartz and opaque minerals	58	17	3	4	3	5	4	3	2	1	<1	1
Granite Fine-grained grey granite, composed of quartz, microcline, oligoclase, biotite, muscovite, accessory minerals, apatite and opaque minerals	76	13	<1	1	<1	<1	3	5	<1	<1	<1	1
Garnetiferous mica-schist Muscovite, biotite, chlorite, garnet, quartz, oligoclase, iron ore and apatite	58	19	1	7	2	2	2	4	1	<1	<1	2
Limestone Calcite	1	<.1	Nil	1	<1	55†	<.1	<1	Nil	<.1	Nil	<.1
Slate Quartz, chlorite, muscovite, apatite, tourmaline, albite and opaque minerals	57	20	10*	3	3	1	3	3	1	<1	Nil	Nil

* Total Fe calculated as Fe_2O_3; † CO_2 = 43%

particularly in the northern hemisphere where many land surfaces have risen since the disappearance of the ice of the glacial period.

Solifluction deposits: crudely stratified deposits on slopes, they are formed in a cold climate as a result of alternating freezing and thawing causing mass movement down the slope.

Volcanic ash: a deposit of fine particles of lava ejected during an eruption. These deposits usually occur very close to the volcano, but if the particles are shot high into the air they may be deposited more than 100 km away.

The most important properties of mineral parent materials are their chemical and mineralogical properties; these are responsible largely for the course of soil formation and the resulting chemical, mineralogical and physical composition of the soil, including the secondary minerals produced by weathering. Two other important properties of mineral parent materials are their permeability and specific surface area.

The organic parent materials which are of restricted distribution are usually composed predominantly of unconsolidated dead and decaying plant remains.

Chemical and mineralogical composition of parent materials

The minerals occurring in rocks such as granite, basalt, gneiss and schist are termed *primary minerals*. In the soil these minerals are decomposed to form *secondary minerals*, particularly the clay minerals. Generally, minerals can be divided into non-silicates and silicates. The non-silicates include oxides, oxyhydroxides, sulphates, chlorides, carbonates and phosphates. Most have relatively simple structures but they vary widely in their solubility and resistance to weathering.

Silicates

The silicate minerals have very complex structures in which the fundamental unit is the silicon-oxygen tetrahedron. This is composed of a central silicon ion surrounded by four closely packed and equally spaced oxygen ions – four-fold coordination. The whole forms a pyramidal structure, the base of which is composed of three oxygen ions with the fourth forming the apex (Figs. 2.2

and 2.3). The four positive charges of Si^{4+} are balanced by four negative charges from the four oxygen ions O^{2-} one from each ion, thus each discrete tetrahedron has four negative charges. The tetrahedra themselves are linked together in a number of different ways forming a variety of distinctive and characteristic patterns which form the basis of the classification of these minerals. Furthermore the type of linkage determines their crystal structure as well as their resistance to weathering.

An important variation in the tetrahedral structure is the substitution of Al for Si. This is known as *isomorphous replacement* which causes an imbalance in the charges within the structure, satisfied by cations such as Na, K, Ca and Mg. Silicates are divided broadly into framework silicates, chain silicates, ortho- and ring silicates and sheet silicates.

Fig. 2.2 Models of silicon-oxygen tetrahedra, (*left*) complete model, (*right*) model with apical ion removed to show the smaller silicon ion

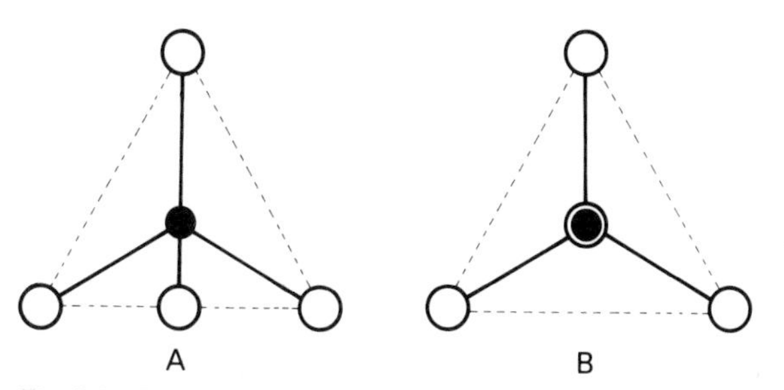

Fig. 2.3 Diagrammatic representation of a silicon-oxygen tetrahedron: (A) vertical; (B) plan

Framework silicates

These minerals are composed of tetrahedra linked through their corners into a continuous three-dimensional structure. The simplest member of this class is quartz which is one of the most common minerals in soils and is composed entirely of silicon-oxygen tetrahedra linked through the oxygen ions so that the resulting structure has twice as many oxygen ions as silicon ions and the formula can be written $(SiO_2)_n$. In rocks it is seen as colourless or milky crystals with conchoidal fracture and is among the minerals most resistant to breakdown, accumulating as one of the main residues during weathering.

The other main group of framework silicates is the colourless, pale pink or white feldspars that are very common in most igneous and metamorphic rocks. They also have a three-dimensional arrangement of tetrahedra but in this case there is a considerable amount of aluminium due to isomorphous replacement and therefore they contain a high proportion of basic cations satisfying the residual charges.

Chain silicates

The two main divisions within this group are the pyroxenes and amphiboles. The pyroxenes occur mainly in basic igneous rocks and range from colourless to dark green with enstatite and hypersthene being good examples. They are composed of tetrahedra linked to each other by sharing two of the three basal corners to form continuous chains (Fig. 2.4) which are linked laterally by various cations such as Ca, Mg, Fe, Na and Al.

The amphiboles exemplified by green hornblende occur in many igneous and metamorphic rocks. They have chains that are double the width of the pyroxene chains and can be regarded as a single band of tetrahedra arranged in a hexagonal pattern (Fig. 2.5). The bands have various dispositions with respect to each other and like the pyroxenes are linked by Ca, Mg, Fe, Na and Al ions.

Ortho- and ring silicates

The greatest variations in the structure of the primary minerals occur in this group which includes the olivines, zircon, titanite

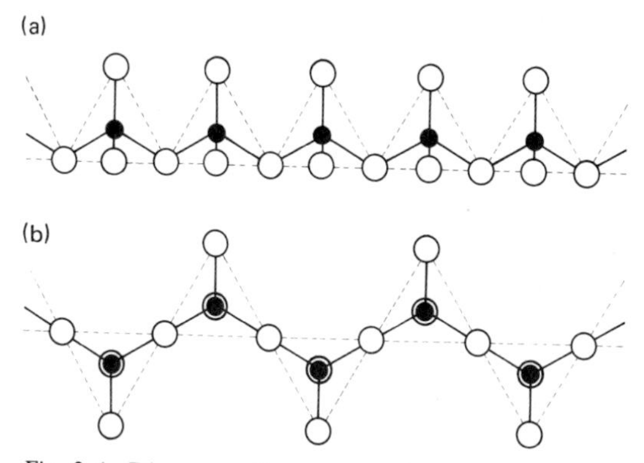

Fig. 2.4 Diagrammatic representation of a pyroxene chain: (a) vertical view and (b) plan view. The points of linkage are through the unsatisfied negative charges of the oxygen ions

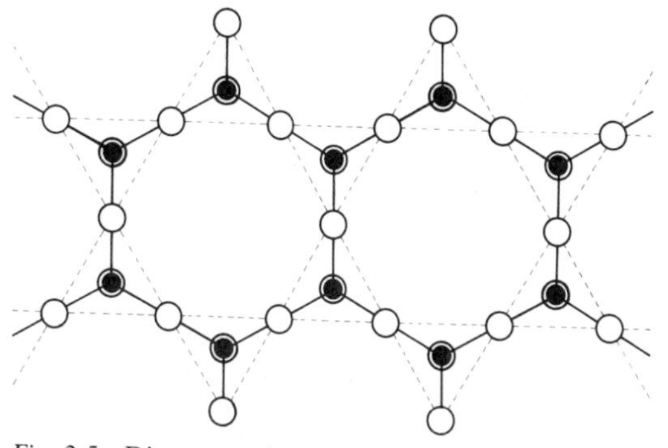

Fig. 2.5 Diagrammatic representation of an amphibole chain, plan view. The vertical view is identical to that of the pyroxene chain. The points of linkage are through the unsatisfied negative charges of the oxygen ions

and garnet. The olivines have a relatively simple structure with separate silicon-oxygen tetrahedra arranged in sheets and linked by Mg and/or Fe. In contrast, there is the complex structure of zircon in which each zirconium is surrounded by eight oxygen ions.

Sheet silicates

This group includes minerals such as muscovite, biotite and the secondary clay minerals. They can be regarded as being composed of various combinations of three basic sheets namely the silicon tetrahedral sheet, the aluminium hydroxide sheet and the magnesium hydroxide sheet.

SILICON TETRAHEDRAL SHEET This is composed of silicon-oxygen tetrahedra linked together in a hexagonal arrangement with the three basal oxygen ions of each tetrahedron in the same plane and all of the apical oxygen ions in a second plane. Thus the silicon tetrahedral sheet is a hexagonal planar pattern of silicon-oxygen tetrahedra (Figs. 2.6 and 2.7).

ALUMINIUM HYDROXIDE SHEET The basic unit of this sheet is the aluminium-hydroxyl octahedron in which each aluminium ion is surrounded by six closely packed hydroxyl groups – six-fold coordination (Fig. 2.8). They are arranged in such a way that there are two planes of hydroxyl ions with a third plane containing aluminium ions sandwiched between the two hydroxyl planes. Note that in order to satisfy all the valencies in the

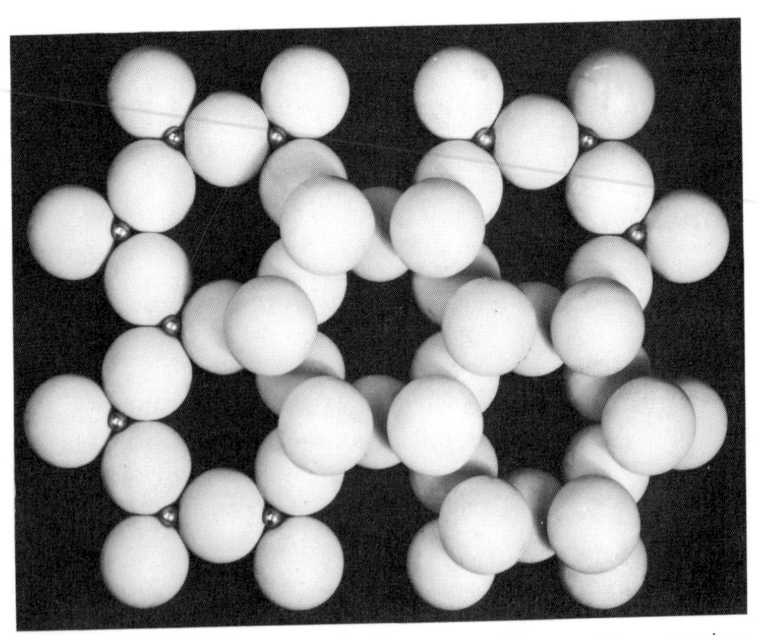

Fig. 2.6 Model of silicon-oxygen tetrahedral sheet with some of the oxygen ions removed to show the smaller silicon ions

structure only two out of every three positions in the aluminium ion plane are occupied by aluminium ions forming a dioctahedral structure.

MAGNESIUM HYDROXIDE SHEET This has a similar structure to the aluminium hydroxide sheet but the aluminium is replaced by magnesium, and because magnesium is divalent all the sites in the middle plane are occupied, forming a trioctahedral structure (Fig. 2.9).

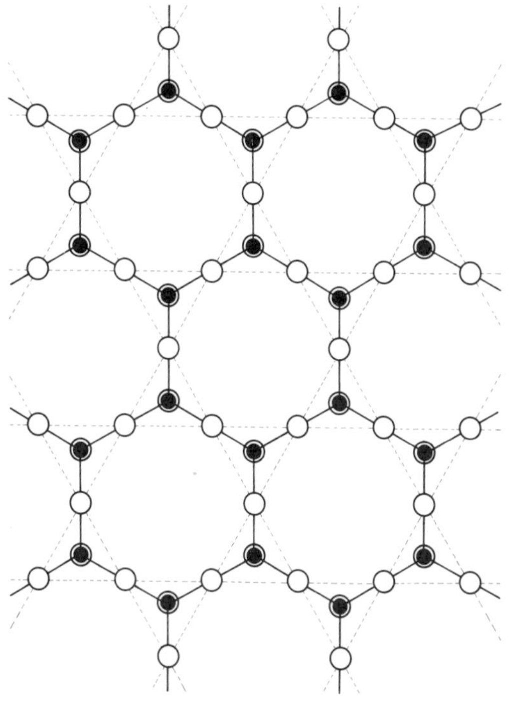

Fig. 2.7 Diagrammatic representation of a silicon-oxygen tetrahedral sheet

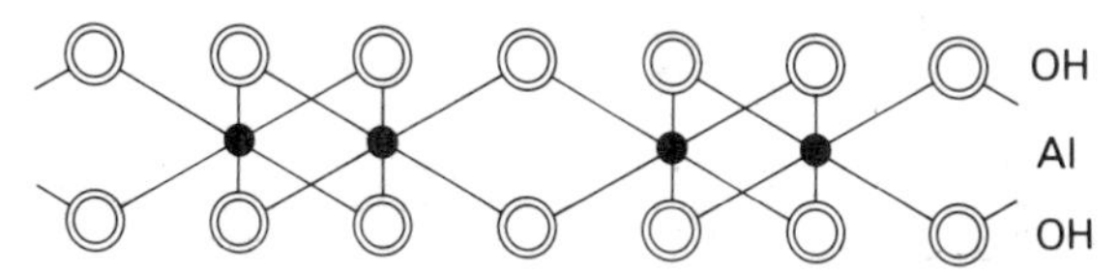

Fig. 2.8 Diagrammatic representation of an aluminium hydroxide sheet – dioctahedral structure

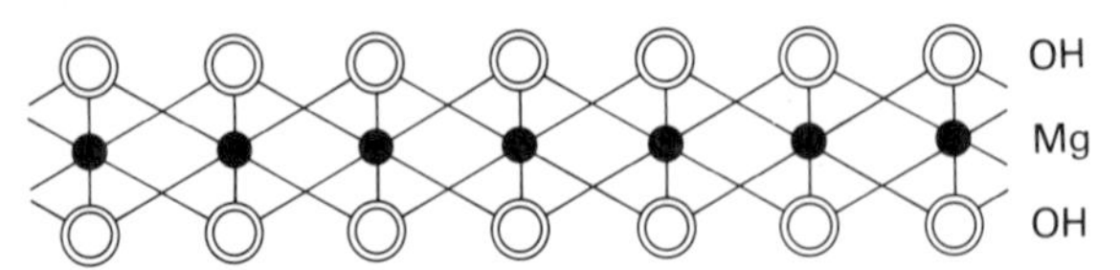

Fig. 2.9 Diagrammatic representation of a magnesium hydroxide sheet – trioctahedral structure

Muscovite

This is composed of one aluminium hydroxide sheet lying between two silicon tetrahedral sheets in which one quarter of the silicon ions have been substituted by aluminium. This causes an imbalance in the structure that is satisfied by potassium.

Biotite

This is composed of one magnesium hydroxide sheet in which there is about one third substitution of iron Fe^{2+} lying between two silicon tetrahedral sheets with one quarter substitution by aluminium. This causes an imbalance in the structure that is satisfied by potassium.

Clay minerals

There are seven types of clay minerals important in soils, namely: kaolinite, halloysite, montmorillonite, hydrous mica, vermiculite, chlorite and allophane. The first six are crystalline and composed of silicon tetrahedral, aluminium hydroxide and magnesium hydroxide sheets in various combinations.

1. *Kaolinite* is the commonest clay mineral in soils and is composed of one aluminium hydroxide sheet and one silicon tetrahedral sheet in which each apical oxygen ion of the silicon tetrahedral sheet replaces one hydroxyl group of the aluminium hydroxide sheet and forms what is known as a 1:1 type of structure (Fig. 2.10). Kaolinite has a well-developed pseudo-hexagonal crystal structure in which the individual crystals range from 0.2–2 μm (Figs 2.11 and 2.12).
2. *Halloysite* has a similar composition to kaolinite but has a tubular form.
3. *Montmorillonite* has a 2:1 type of structure in which one

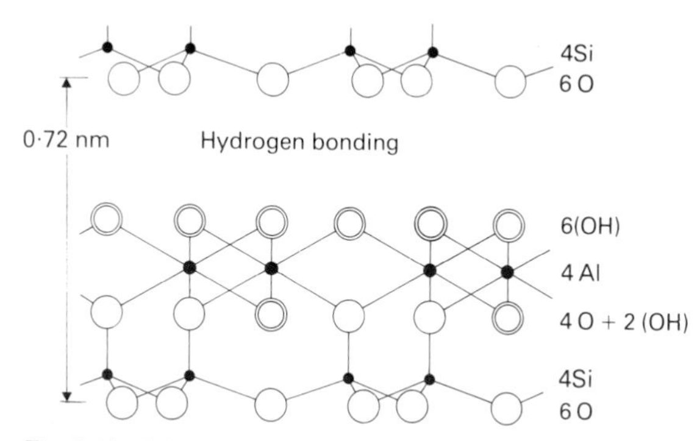

Fig. 2.10 Diagrammatic representation of the structure of kaolinite

Fig. 2.11 Transmission electron photomicrograph of kaolinite

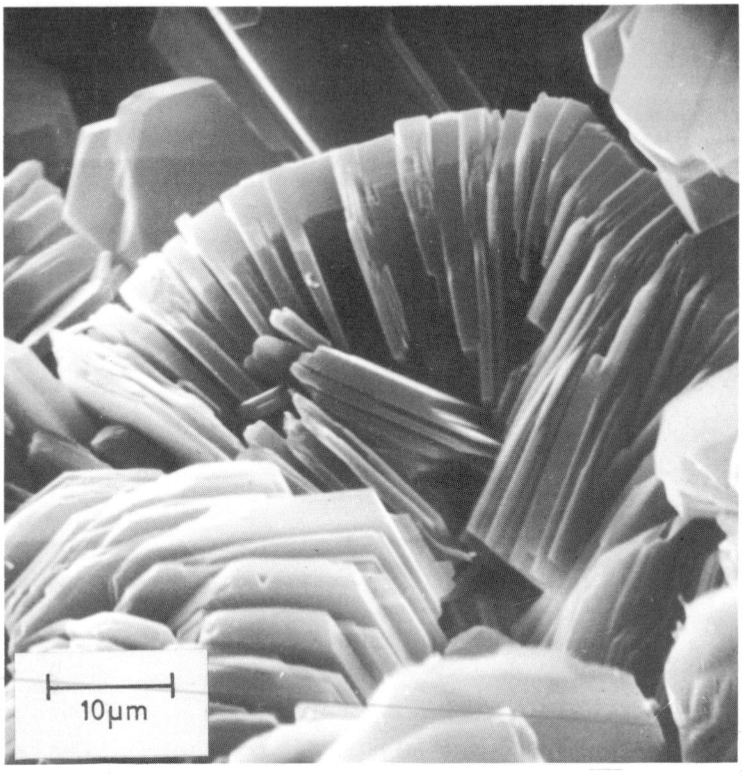

Fig. 2.12 Scanning electron photomicrograph of vermiform kaolinite composed of well-developed hexagonal plates

aluminium hydroxide sheet with substitutions of iron and magnesium lies between two silicon tetrahedral sheets with some aluminium substitution (Fig. 2.13). This mineral has a very small particle size and a great affinity for water. This imparts one of the most important properties to this mineral – namely its ability to expand and contract in response to the addition or loss of water.

4. *Hydrous mica* can be regarded as clay size particles of muscovite and biotite.
5. *Vermiculite* can be regarded as hydrated mica from which the potassium has been removed.
6. *Chlorite* is a 2:1 layered silicate composed of a magnesium hydroxide sheet sandwiched between two mica layers.

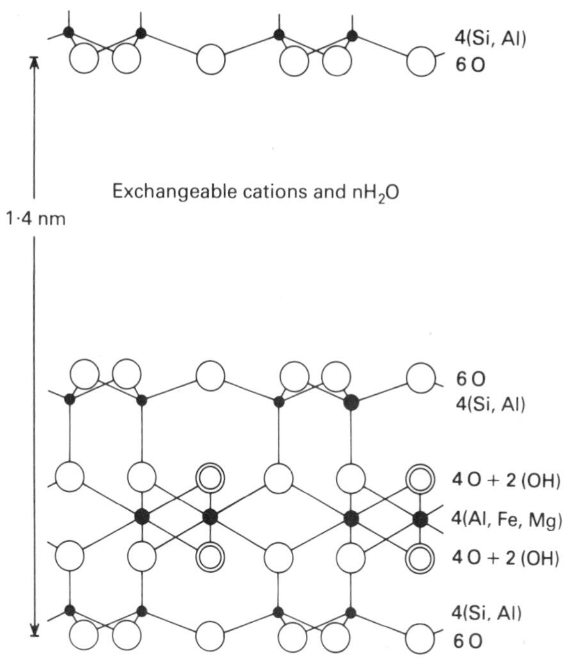

Fig. 2.13 Diagrammatic representation of the structure of montmorillonite

7. *Allophane* is amorphous or very finely crystalline with an indeterminate composition but generally has about equal amounts of aluminium hydroxide and silica.

The upper 10 km of the Earth's crust is made up predominantly of the elements shown in Table 2.2. Oxygen occupies 90 per cent by volume and is the most abundant element but it usually occurs in combination with other elements – particularly with silicon, mainly within the more complex silicate structures. Thus the majority of the resulting soils are composed predominantly of silica, either as quartz or in a combined form in silicates. The third most frequent element is aluminium which is found mainly in feldspars and in sheet silicates but which occurs also in varying proportions in the other silicates. Iron is also of fairly widespread distribution, occurring in large amounts in relatively few minerals of which biotite, some pyroxenes, amphiboles and olivines are the principal contributors and it should be noted that it is mainly in the ferrous state. Sodium and potassium are found chiefly in

Table 2.2 Distribution of elements in the Earth's crust

Element	*Per cent (by wt.)*	*Element*	*Per cent (by wt.)*
Aluminium	8	Oxygen	47
Calcium	4	Phosphorus	<1
Carbon	<1	Potassium	3
Chlorine	<1	Silicon	28
Iron	5	Sodium	3
Magnesium	2	Sulphur	<1
Manganese	<1	Titanium	<1

the feldspars but significant amounts of potassium also occur in the micas. Calcium and magnesium have a wide distribution among the silicates as well as being the principal cations in many non-silicates such as calcite and dolomite which are the dominant minerals in limestone. Phosphorus is very restricted in its distribution, occurring only in significant proportions in apatite which is fairly widespread in small amounts in parent materials but the total content of this mineral in soils seldom exceeds 0.2 per cent. There are also a number of distinctive minerals such as tourmaline, zircon and rutile that occur in low frequencies and are known as *accessory minerals*. Some of these minerals contribute the vital micro-elements (see page 137).

While most of the silicates originate in igneous and metamorphic rocks, the clay minerals are formed within soils or are inherited from parent materials such as lacustrine deposits and shale.

Surface area of parent materials

The specific surface area of the constituent particles in the parent material determines the amount of interaction that is possible with the environment, particularly with water. Consolidated rocks have an extremely small surface area when compared with alluvial sands which in turn have a smaller area than clays, thus surface area increases as particle size decreases. Variations in surface area and particle size distribution have a profound effect on the speed of soil formation for it is found that soils will develop in sediments much faster than from consolidated rock of the same composition.

Permeability of parent materials

The permeability of the parent material influences the rate of moisture movement which in turn influences the speed of soil formation. The most permeable materials and therefore those that allow free movement of moisture usually have a high content of sand but as the particle size decreases the general tendency is for the material to become more impermeable so that most clay parent materials allow only a very slow rate of moisture movement.

Perhaps compaction should be mentioned as a property of parent materials since some superficial deposits may be compact and will restrict moisture movement. Thus a sandy deposit may have low permeability because of compaction. Another property that influences permeability is structure or the degree of organisation of the soil, as discussed on page 94. This is particularly important in connection with drainage and crop growth.

Classification of parent materials

The classification is based on the easily recognisable intrinsic characteristics and uses a system of letter symbols for designating the parent material in the field and in written descriptions. There are nine classes, five are based on the proportion of ferromagnesian minerals, one on the amount of carbonate, the seventh and eighth on the amounts of salts and the ninth on the amounts of organic matter. Set out in Fig. 2.14 are the nine classes. Each of the classes is subdivided into consolidated and unconsolidated materials.

Consolidated materials

These are sub-divided into four types:

1. *Non-crystalline*: composed of a continuous phase without crystals, e.g. volcanic glass.

Type	Composition		Symbol
Ultrabasic	>90% ferromags		U
Basic	40–90% ferromags	45–55% SiO_2	B
Intermediate	20–40% ferromags	55–65% SiO_2	I
Acid	5–20% ferromags	65–85% SiO_2	A
Extremely acid	<5% ferromags	—	E
Carbonate			C
low carbonate	1–5% $(Ca+Mg)CO_3$		1C
medium carbonate	5–20% $(Ca+Mg)CO_3$		2C
high carbonate	20–50% $(Ca+Mg)CO_3$		3C
dominant carbonate	>50% $(Ca+Mg)CO_3$		4C

Type	Composition	Symbol
Sulphate		S
slightly sulphate	1–5% $CaSO_4$	1S
moderately sulphate	5–20% $CaSO_4$	2S
strongly sulphate	20–50% $CaSO_4$	3S
dominantly sulphate	>50% $CaSO_4$	4S
Saline		H
slightly saline	1–5% soluble salts	1H
moderately saline	5–20% soluble salts	2H
strongly saline	20–50% soluble salts	3H
dominantly saline	>50% soluble salts	4H
Organic		P
slightly organic	1–5% organic matter	1P
moderately organic	5–20% organic matter	2P
strongly organic	20–50% organic matter	3P
dominantly organic	>50% organic matter	4P

CONSOLIDATED

Mineral size		Symbol
Non-crystalline		N
Fine grained	<1 mm	F
Medium grained	1–5 mm>	M
Coarse grained	<5 mm	K

UNCONSOLIDATED

*Particles < 2 mm			
Name	Symbol	Name	Symbol
Clay	c-	Sandy clay	sc
Silty clay	zc	Sandy clay loam	scl
Silty clay loam	zcl	Loam	1-
Silt	z-	Sandy loam	sl
Silt loam	zl	Loamy sand	ls
Clay loam	cl	Sand	s-

*Particles > 2 mm	
Name	Symbol
Gravelly	G
Stony	S
Very stony	VS
Bouldery	B

* Same as for texture

Fig. 2.14 Classification of parent materials

2. *Fine grained*: material containing minerals that are just visible with the naked eye but there may be a few larger crystals (phenocrysts).
3. *Medium grained*: these rocks are composed of minerals up to 5 mm in diameter.
4. *Coarse grained*: rocks with minerals greater than 5 mm.

Unconsolidated materials

There are two principal subdivisions: the <2 mm and >2 mm fractions. The <2 mm fraction is further subdivided into twelve particle-size classes and the >2 mm fraction into four size classes.

By combining the symbols given in Fig. 2.14 any parent material can be given a designation indicating its mineralogy, degree of consolidation, particle size distribution and degree of stoniness. A few examples are:

AK: acid, coarse-grained crystalline rock such as granite.
BF: basic, fine-grained rock such as basalt.
2Cz-: moderately calcareous silt.
Ascl: acid sandy clay loam.
EsVs: extremely acid sand, very stony.

Climate

Climate is the principal factor governing the type and rate of soil formation as well as being the main agent determining the distribution of vegetation. The climate of a place is a description of the prevailing atmospheric conditions and for simplicity it is defined in terms of the averages of its components, the two most important being temperature and precipitation. Although averages are most commonly used, diurnal and annual patterns and extremes are not ignored since they give character and sometimes are important factors. For example, the occurrence of occasional high winds can determine the development policy of an area. However, it must be stressed that the atmospheric climatic data do not always give a true picture of the soil climate. For example, the amount of water in the soil may vary considerably within a distance of a few metres from permanently saturated to dry and quite freely draining, whereas there is virtually no difference between the amounts of precipitation at the two sites – one site may be in a depression where moisture can accumulate, and the other on an adjacent slightly elevated situation. Regularly these differences in the moisture regime at the two sites lead to the development of different soils and contrasting plant communities, with a dry habitat community on the elevated situation and a marsh community in the depression.

Temperature

Atmospheric and soil temperature variations are the most important manifestations of the solar energy reaching the surface of the Earth, part of which is absorbed and converted into heat in the atmosphere and soil while the remainder is reflected back. The amount absorbed is influenced by the colour of the soil; since dark coloured soils absorb the most radiation they are the warmest. A proportion of the heat produced is maintained in the soil but some is lost to the atmosphere by convection of hot air from the soil and by back radiation. Also a considerable amount is used for evaporation of moisture into the atmosphere (Fig. 2.15).

Cloudiness, humidity, dust particles and pollution absorb radiation thereby reducing the amount reaching the Earth's surface.

Vegetation has a buffering effect on soil temperature by absorbing and reflecting radiation during the day but during the night it reflects back to the soil some of the heat lost by radiation, so that temperature fluctuations are less beneath forests than in adjacent exposed sites. In a similar way a cover of snow reduces the loss of heat from the soil and may prevent the penetration of frost. On the other hand snow reflects over 90 per cent of the incoming radiation.

Perhaps it should be mentioned that only about 0.1 per cent of all energy reaching the Earth's surface is absorbed by plants and fixed by photosynthesis and is therefore the total amount of energy that runs all the life on our planet.

About half of that energy is used for respiration the other half enters one of two food chains – the decay food chain and the grazing food chain.

The main effect of temperature on soils is to influence the rate of soil formation, since for every 10°C rise in temperature the

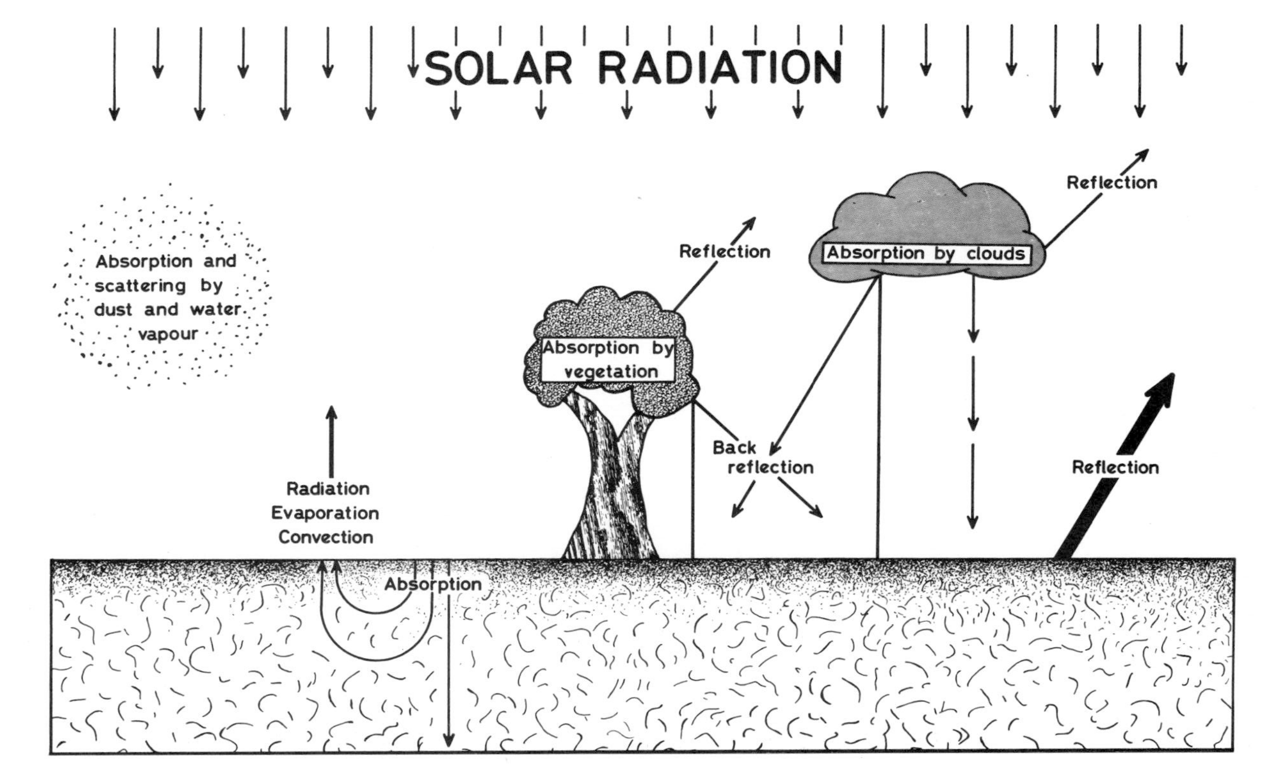

Fig. 2.15 Utilisation of the solar radiation reaching the Earth's surface

speed of a chemical reaction increases by a factor of two to three. The principal process in the soil to which this applies is the weathering of minerals. The rate of both biological activity within the soil and the breakdown of organic matter are also increased by a rise in temperature. In addition the amount of moisture evaporating from the soils is increased.

Development of vegetation can also be affected by soil temperature – in cool climates plants become active at 5°C and reach maximum activity at 20°C, the rate of maturation having increased three to four times.

The amount of radiation reaching the surface and soil temperatures are determined by a number of factors, probably the diurnal and seasonal variations are the most important for they give the soil well marked cycles. During the diurnal cycle in tropical and subtropical areas it is normal for heat to move downwards in the soil during the day from the surface, due to warming by incoming radiation and upwards during the night as the surface cools. This takes place also during the summer period in the middle and higher latitudes, but in these areas during the winter the atmosphere is generally cooler than the soil and the incoming radiation is not sufficient to heat the soil which steadily cools and eventually may freeze from the surface downwards. The widest temperature fluctuations take place at the surface particularly in certain desert areas so that the surface shows both the absolute maximum and absolute minimum temperatures (Fig. 2.16).

Heat moves very slowly down through the soil so that the diurnal maximum in the lower horizons at about 20–30 cm occurs up to twelve hours after the surface maximum (Fig. 2.16). This lag is greater in the annual cycle when the lower horizons attain their maximum even after the surface begins to cool in response to a seasonal change. The temperature fluctuations within the soil between seasons is greater at the surface than in the lower horizons but remains uniform at 10–20 m from the surface (Fig. 2.17). Thus during the summer in middle latitudes the diurnal mean surface temperature is higher than that in underlying layers but in winter the reverse is true.

Aspect and latitude influence soil temperature. Land surfaces normal to the rays of the sun are warmest but the greater the distance from the equator the smaller the amount of radiation reaching the Earth's surface (Fig. 2.18).

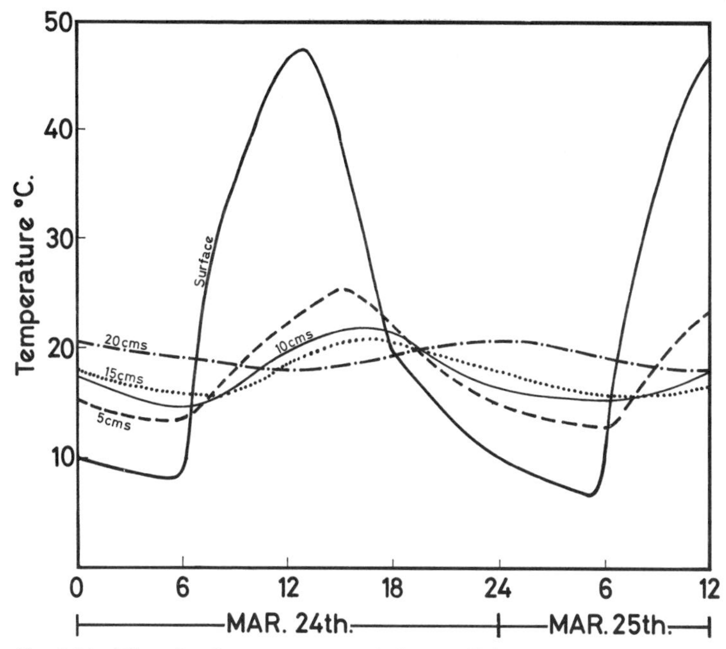

Fig. 2.16 Diurnal soil temperature variations at Cairo

With increasing altitude, the temperature decreases at the rate of about 1°C for each 170 m: on the other hand precipitation increases initially and then decreases. These two factors combine to produce a vertical zonation of climate, vegetation and soil as seen in all mountainous areas.

A classification of soil temperature proposed by the USDA (1975) is given in Table 2.3 and is of value both in land use and soil classification.

Moisture

The differentiation of horizons is determined very largely by the movement of moisture, therefore this process is of paramount importance. In fact, the soil solution might be regarded as the main "conveyor belt" in soils whereby ions and small particles are translocated from one place to another.

The moisture entering the soil is derived mainly from precipitation as rain and snow and contains appreciable amounts

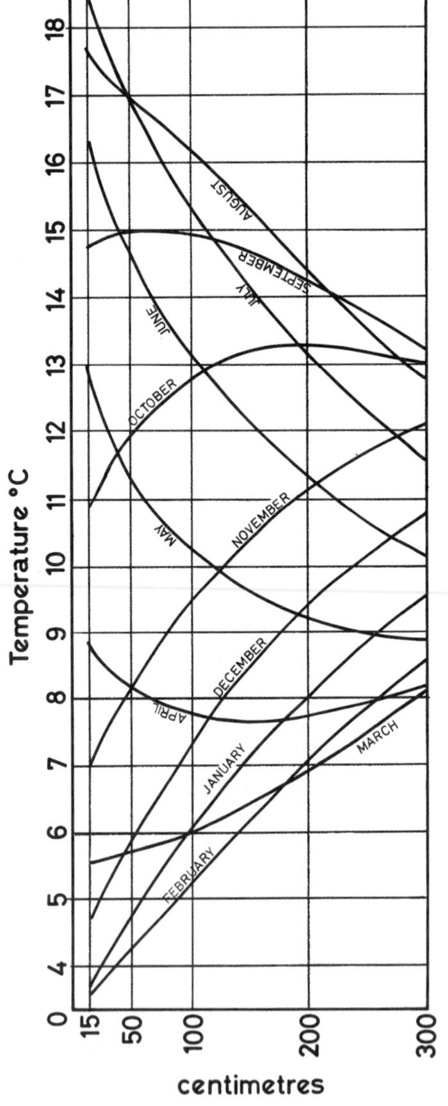

Fig. 2.17 Annual soil temperature variations in southern England

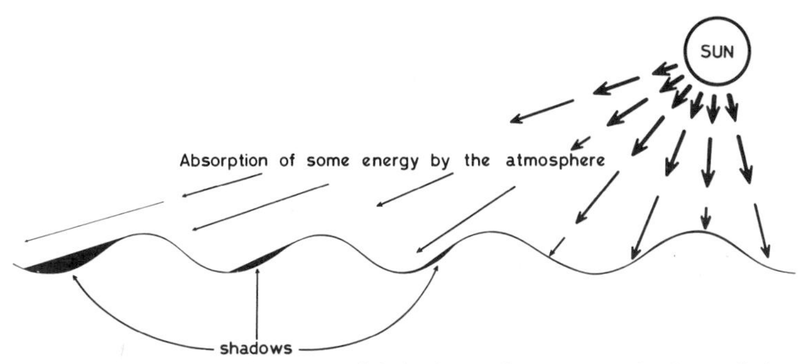

Fig. 2.18 The effect of aspect and latitude on the amount of solar radiation reaching the Earth's surface. The intensity of the radiation is roughly indicated by the thickness of the arrows

Table 2.3 Soil temperature categories

	Mean annual soil temperature	*Difference between mean winter and summer temperature*
Pergelic	<0°C	—
Cryic	>0° but <8°C	—
Frigid	<8°C	>5°C
Mesic	⩾8°C but <15°C	>5°C
Thermic	⩾15°C but <22°C	>5°C
Hyperthermic	⩾22°C	>5°C
Isofrigid	<8°C	—
Isomesic	⩾8°C but <15°C	—
Isothermic	⩾15°C but <22°C	—
Isohyperthermic	>22°C	—

of dissolved CO_2. Thus it is probably more correct to think of moisture entering the soil as a dilute weak acid solution which is much more reactive than pure water. Some moisture takes part in a number of chemical reactions in the soil and some is retained, but by far the greatest amount is lost through drainage or by evapotranspiration, i.e. the combined processes of evaporation and transpiration (Figs. 2.19 and 2.20.).

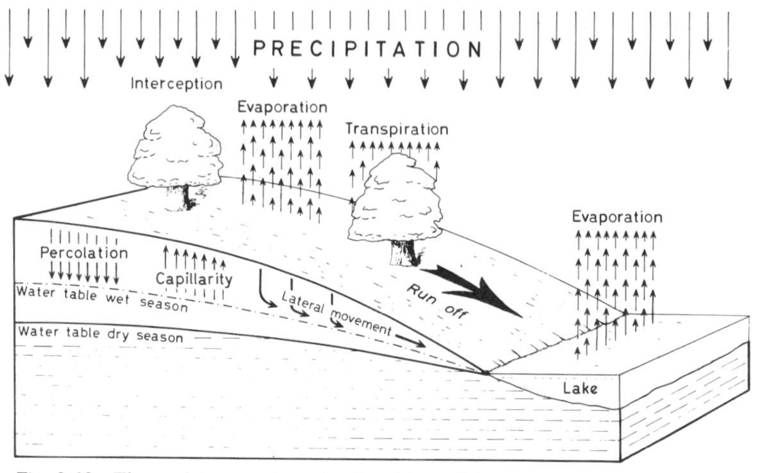

Fig. 2.19 The moisture cycle under humid conditions

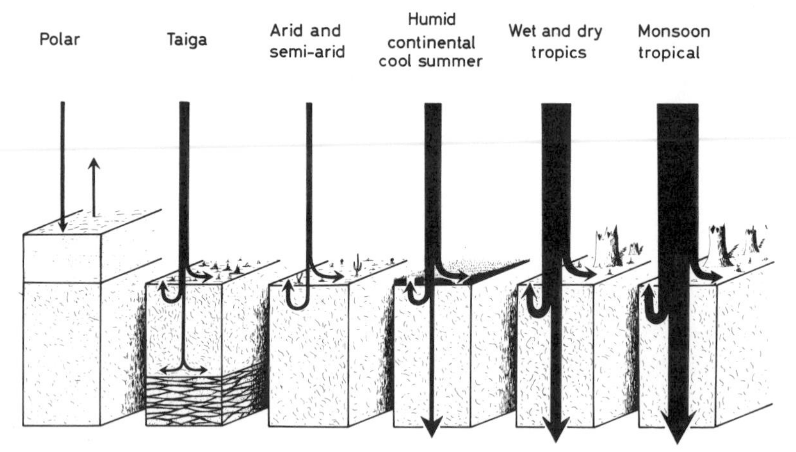

Fig. 2.20 The fate of moisture falling on the surface under a number of different climatic conditions

The intensity of precipitation varies from place to place over the Earth's surface and in areas of bare soil it is precipitation of moderate intensity that is most effective in entering the soil. Light showers of rain that hardly enter the soil are quickly lost by evaporation. Heavy showers may cause moisture to accumulate at the surface particularly on clay soils leading to run-off and creating an erosion hazard. Precipitation can be intercepted by the foliage of vegetation and is later lost by evaporation.

Generally, moisture moves downwards after rainfall or melting snow, but it can move upwards by capillarity in response to drying out at the surface or it may move laterally through the soil on slopes. Moisture movement within soils can also take place in the vapour phase but this seems to be minimal (see page 36 *et seq.*).

One of the most fascinating aspects of climate is its continuous change with time as revealed by an examination of the wide range of geological strata. With regard to soils, the climate at any one point on the surface of the Earth has changed many times during the last two million years, leading to changes in the processes of soil formation, so that most soils have been subjected to contrasting processes as determined by climatic changes (see pp. 54 and 197).

Water characteristics of soils

The state of water in soils varies from that which is free to flow to that which is adsorbed firmly on the surfaces of particles – hygroscopic water. Thus, water is held with varying degrees of tension or suction.

The phenomenon of suction can be illustrated by placing a drop of water onto a dry soil aggregate. Almost immediately the water is drawn into the soil. After the initial intake it will continue to spread and then stop when the films of water around the grains become very thin and are held firmly or with great tension on their surfaces. The usual way of expressing tension is that of the bar which is the pressure exerted by a column of water 1023 cm in height and is approximately the standard atmospheric pressure. The second unit is the millibar (mbar) or 1/1000 bar. Thus the suction of a 1000 cm column of water is about 1 bar, a 100 cm column is 1/10 bar (100 mbar) and a 10 cm column of water about 1/100 bar (10 mbar).

Two important moisture characteristics are the field capacity and the wilting point. After the soil has been saturated and the excess water drained away the soil is said to be at *field capacity*. If plants are growing on the soil they will extract moisture until they cannot extract any more then they will wilt and die if the soil is not rewetted. The point at which permanent wilting starts is known as the *wilting point*. The actual amount varies much from soil to soil and plant to plant. Thus water held between field capacity and wilting point is the water available to plants. The amount varies considerably from soil to soil as shown in Figs. 2.21 and 2.22. A significant feature is that much of the water held by clay soils beneath field capacity is also beneath the wilting point.

Water movement in soil

There are three different types of movement depending upon the amount of water present. These are saturated flow, unsaturated flow and vapour transfer.

SATURATED FLOW This takes place when water is moving through the soil in which all of the pores are filled with water. This occurs most commonly beneath the water table but also whenever the soil becomes saturated. Saturated flow may be in any direction – vertically downwards or laterally over a pan. The precise rate of flow is determined by the hydraulic conductivity, i.e. the ease with which water will pass through the soil as determined by the size and distribution of the pores.

UNSATURATED FLOW Under these conditions water moves from pore to pore by flowing over the surfaces of aggregates and/or particles while at the same time there is a considerable amount of air in the large pore spaces. In soils that have recently been wetted the flow is downwards in response to gravity but after field capacity has been attained movement is usually either lateral or upwards by capillarity in response to a moisture gradient due to drying at the surface or uptake by plant roots. Thus unsaturated flow takes place in response to gravity as well as in response to a moisture gradient. It is the main process responsible for translocation of material in solution and suspension and also tries to equilibrate the moisture tension in soil.

VAPOUR TRANSFER This is the movement of water in the vapour phase from one place to another in the soil and from the soil to

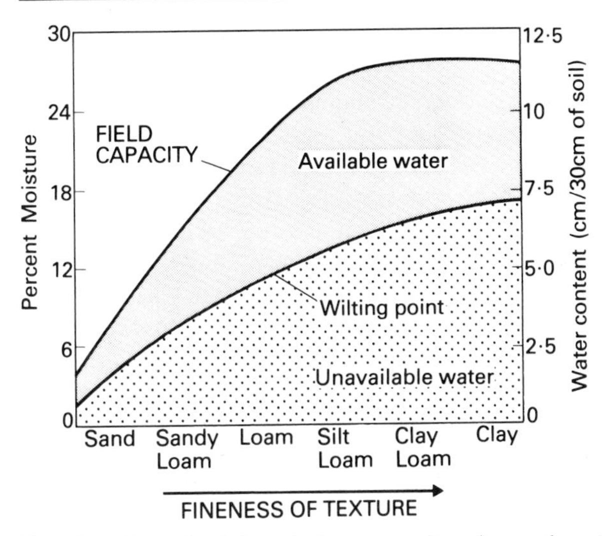

Fig. 2.21 General relationship between soil moisture characteristics and soil texture. Note that the wilting point increases as the texture become finer. The field capacity increases up to silt loams then levels off

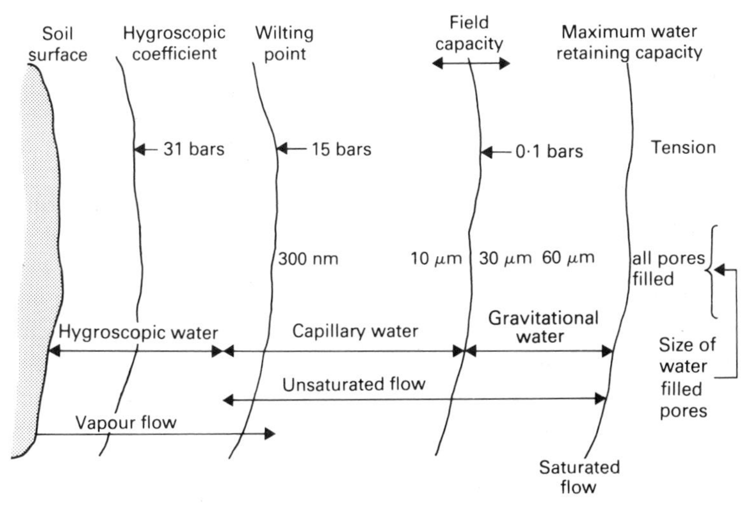

Fig. 2.22 Relationships between the types of water, types of flow and suction

the atmosphere. The rate of movement is determined by the relative humidity, the temperature gradient, the size and continuity of the pores and the amount of water present. The humidity of the above ground atmosphere very largely determines the amount of water that is lost from the soil affecting both evaporation and transpiration. The temperature gradient is most important at the surface which, as shown on page 30, may have very high temperatures causing a large amount of vaporisation of water. The size and continuity of the pores will determine whether there can be free movement from place to place within the soil. In order to understand more fully the movement of water in the soil-plant system it is necessary to consider the following situations.

Water movement in coarse sand

Consider the situation in which water is continuously added to the surface. It will flow freely downwards by saturated flow in response to gravity. If the addition is stopped, downward saturated flow will soon change to unsaturated flow because there is insufficient water to fill the pores, there is only enough to flow over the surfaces of the grains. Eventually downward flow will almost cease when the upper part of the sand reaches field capacity. At the surface, loss of water by evaporation creates a moisture gradient causing water to be drawn upwards by capillarity.

Water movement in a clay

In a saturated clay, water movement is extremely slow and, through drainage, scarcely ever takes place. Thus the concept of field capacity is not really applicable: the available water will range from wilting point to the point of anaerobism that will inhibit root growth. Clay soils lose water mainly by evapotranspiration. The individual particles lose their films of water and are drawn closer and closer together. This causes shrinkage and ultimately the development of a crack system. With the onset of the wet season water flows rapidly down the cracks, rewetting the soil from the bottom upwards.

Another interesting feature about soil–water relationships is that for a given amount of precipitation a sandy soil is more leached than a clay soil and there are situations where there is sufficient rainfall to bring a sandy soil – but not a clayey soil – to field capacity.

Organisms

Nearly every organism living on the surface of the earth or in the soil affects the development of soils in one way or another. The organisms can be considered under the following headings:

Higher plants	Microorganisms
Vertebrates	Mesofauna

Higher plants

Higher plants influence the soil in many ways. By extending their roots into the soil they act as binders and so prevent erosion from taking place, with grasses being particularly effective in this role. Roots can also grow within cracks in rocks forcing them apart. When plants die and their roots decay, they leave a network of passages through which water and air can circulate more freely.

One of the greatest contributions of the higher plants is through the addition of organic matter or litter to the surface. The total amount added by the different plant communities is very variable but it is no guide to the amount present in the soil which depends more upon the rate and type of breakdown. Tropical plant communities contribute as much as 25 tonnes ha^{-1} a^{-1}, tall grass prairie 5.0 tonnes and pine forest 2.5 tonnes but the amount of organic matter in the soils beneath these communities is in the reverse order. Pine forests may have an accumulation of about 15 cm of organic matter at the surface. Prairie grass soils have up to 15 per cent organic matter incorporated in the mineral soil while the soils of tropical rain-forests often contain less than 5 per cent organic matter. In comparison dairy cows produce about 3.5–4.5 tonnes ha^{-1} a^{-1} and dead roots produce a third to a half of the leaf fall.

Plants extract water and nutrients from the body of the soil and under natural conditions return most of the nutrients to the surface in their litter which decomposes and releases them, rendering them available for re-absorption (see page 137). The type of plant community can sometimes be used to assess soil conditions. Rushes usually indicate wet soils while heather grows best on dry acid soils.

Vertebrates

A few mammals, including rabbits, moles and the prairie dog, burrow deeply into the soil causing considerable mixing, often bringing subsoil to the surface. The classical examples of this are the crotovinas found in many Chernozems, especially those of Europe where the blind mole rat is principally responsible (Fig. 2.23 and Colour Plate IIA).

Uncontrolled grazing by animals such as goats will devour the vegetation and leave the surface bare for erosion. This is a conspicuous feature in many countries bordering the Mediterranean Sea and many semi-arid areas (Figs. 6.6, 6.7 and 6.8).

Microorganisms

The predominant microorganisms are bacteria, fungi actinomycetes, algae and viruses. Bacteria are rod shaped and

Fig. 2.23 Mole hills in a field of cultivated grass. Each mole hill is about 30 cm in diameter and 15 cm high

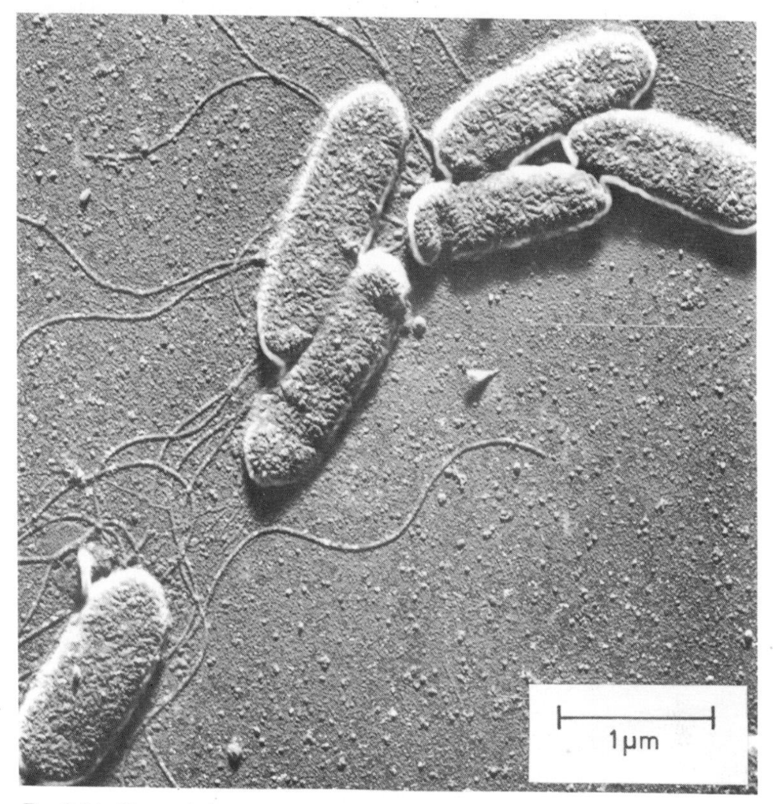

Fig. 2.24 Transmission electron photomicrograph of carbon replicas of flagellated manganese-oxidising bacteria – *Pseudomonas* sp. The thread-like flagella help to propel the bacteria in water

about 1 μm in length (Fig. 2.24). They are the smallest and most numerous of the free-living microorganisms in the soil, where they number several million per gram with a live weight of 1000–6000 kg ha^{-1} in the top 15 cm. This weight is slightly less than that of the fungi but greater than that of the other microorganisms combined.

The actinomycetes are second in abundance to bacteria preferring dry warm grassland and neutral conditions. There are a large number of genera of which the *Streptomycetes* are dominant. They have a characteristic musty odour and produce antibiotics and enzymes that kill bacteria and other

Fig. 2.25 Scanning electron photomicrograph of fungal hyphae on a soil surface – the inset is an enlargement of a fruiting body

microorganisms. They are also very important as decomposers of organic matter particularly the most common polysaccharides and chitin.

The algae are early colonisers of newly exposed material in wet situations such as paddy fields and the very widespread shallow pools in the arctic. When in sufficient numbers they help to form a crust at the soil surface thereby preventing soil erosion. Also they make an important early contribution to the organic matter in the soil and are thus early initiators of the carbon and nitrogen cycles.

Viruses are generally regarded as parasites in larger animals and plants but nearly every class of microorganism in the soil is subject to viral attack. Each individual virus has a limited host range, some attacking bacteria and other types attacking fungi,

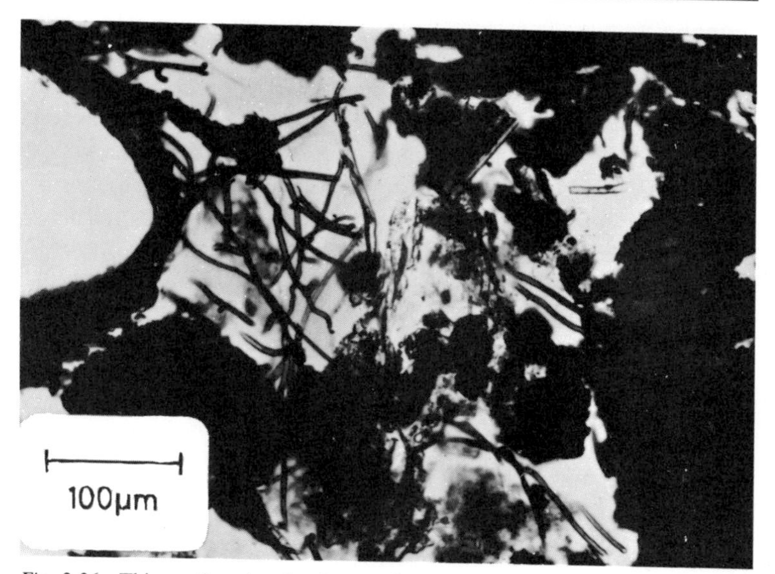

Fig. 2.26 Thin section showing fungus decomposing organic matter

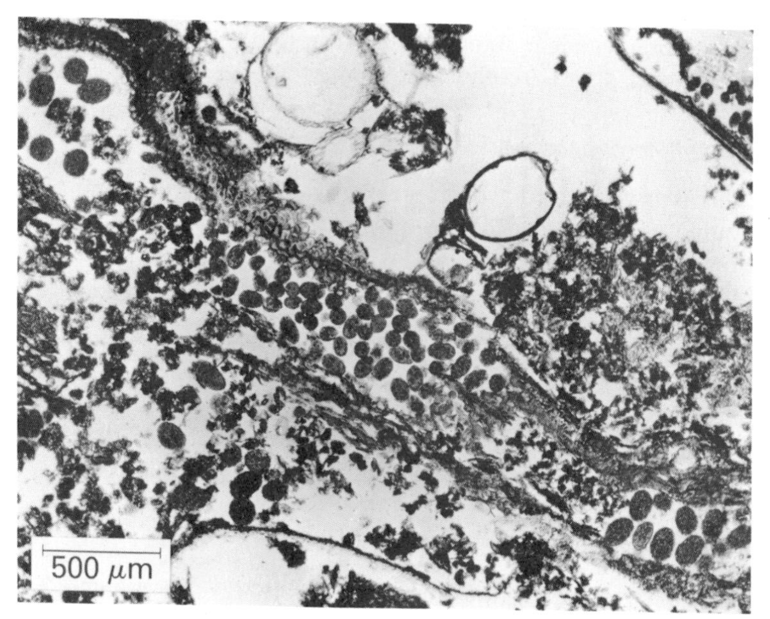

Fig. 2.27 Faecal pellets of mites

actinomycetes, algae, protozoa, earthworms, etc. The virus attacking bacteria or bacteriophage has a head and tail with a diameter of 0.05–0.01 μm. The virus enters through the cell wall of the bacterium and attaches itself by its tail then causes lysis – decomposition of the cell walls. Viruses specific to Azotobacter, Rhizobium and Nitrobacter are known (see page 69).

The distribution of microorganisms in soils is determined largely by the presence of a food supply, therefore they occur in the greatest numbers in surface horizons which have a teeming mass of biological activity. Most members require an aerobic environment and have optimum temperature requirements of 25–30°C. It should be mentioned that these high temperatures occur in only a few soils, therefore microorganisms usually operate below their optimum.

There are four classes of microorganisms based on their sources of carbon and energy.

Photoautotrophs

The organisms contain chlorophyll and utilise light as their energy source and CO_2 as their principal source of carbon; they include the blue-green algae some of which can fix atmospheric nitrogen. The higher plants are photoautotrophs.

Photoheterotrophs

This is a very restricted group of organisms that use light as a source of energy and derive much of their carbon from organic compounds.

Chemoautotrophs

These organisms derive their energy from the oxidation of inorganic compounds and use CO_2 as their principal source of carbon. They include several groups of specialised bacteria, including the all important nitrifying bacteria (see page 68).

Chemoheterotrophs

This is by far the largest group of microorganisms that utilise organic compounds both as a source of energy and carbon. They include protozoa, fungi, actinomycetes and most bacteria and are of immense importance through their participation in humification and ammonification (see page 68).

Within the soil there are many interactions between the microorganisms. In some cases one organism depends directly upon the activity of another. A good example is the production by *Nitrosomonas* of nitrite which acts as a substrate for *Nitrobacter*. Other examples are mycorrhizae and root nodules. A mycorrhiza is a composite fungus – root organ which occurs on many conifers. Mycorrhizae are of two types, ectotrophic and endotrophic. The former develop complete sheaths of fungal tissue which enclose the root tips while the latter which are more widespread ramify through the cortex of the roots. They behave like root hairs by taking up nutrients and water.

Root nodules are formed by the bacterium *Rhizobium* sp. beneath the epidermis of the roots of some plants particularly members of the *Leguminosae*. The Rhizobium fixes atmospheric nitrogen which passes into the plant as a nutrient.

Microorganisms will decompose most carbonaceous materials in soils including pollutants. This helps to reduce the build-up of pesticides and herbicides and organic industrial wastes. Initially pesticides may cause a decrease in the microbial population but gradually resistant strains develop.

Mesofauna

This group includes earthworms, enchytraeid worms, nematodes, mites, springtails, millipedes, some gastropods and many insects, particularly termites and ants. Like microorganisms their distribution is determined almost entirely by their food supply and therefore they are concentrated in the top 2 to 5 cm; only a few, such as earthworms penetrate below 10 to 20 cm. Generally the mesofauna require an aerobic environment with conditions around neutrality but many can live in either acid or alkaline soils. The concentration of each organism varies greatly from place to place, but it is estimated that under optimum conditions the biomass of earthworms is about 800 kg ha^{-1} and nematodes about 5–20 million/m^2.

The mesofauna are concerned largely with the ingestion and

decomposition of organic matter (Fig. 2.27), in addition many earthworms, larvae and millipedes ingest both mineral and organic matter and as a consequence they produce faecal material which is a homogeneous blend of these two substances. They also transport material from one place to another and in so doing they produce passages which improve drainage and aeration. (Figs. 2.28, 2.29 and 2.30). Earthworms generally transport material to the surface while termites transport material to build their termitaria. In some situations harvester ants will denude an area 2–6 m around their nest. Some mesofauna, particularly the nematodes, transmit virus diseases or are themselves parasites of great economic importance such as the potato eelworm.

Topography

This includes the dramatic mountain ranges and the flat featureless plains, both of which give the impression of considerable stability and seem to be timeless. However, this is not the case, for it is known from many investigations that all land surfaces, even those in areas of hard rocks such as granite, are constantly changing through weathering and erosion. Thus topography is not static but forms a dynamic system, the study of which is known as *geomorphology*.

Topography influences the soil in many ways. For example, the thickness of the soil is often determined by the nature of the relief. On flat or gently sloping sites there is the tendency for material to remain in place and for the soil to be thick but as the slope increases so does the erosion hazard, resulting in thin stony soils on strongly sloping ground. Topography also influences the drainage and the amount of moisture in the soil (Fig. 2.31). Although landscapes may seem to be very complex most can be considered as having four elements as shown in Fig. 2.32.

Topographic features are produced by three main processes, tectonic processes (crustal disturbances), erosion and deposition. Initially all topographic features are produced by tectonic processes; the surfaces are then acted upon by running water, ice, frost and wind which are the principal agents of erosion and deposition.

The material removed by running water is normally

Fig. 2.28 A termitarium in western Nigeria

transported to rivers to form alluvial deposits or taken to the sea to form deltas.

Erosion by ice in the form of glaciers is relatively local at present but was more widespread in the past when they produced characteristic erosional forms such as U-shaped valleys and depositional features such as drumlins and moraines that are

Fig. 2.29 Soil profile with a small termitarium near the bottom

common over large parts of Europe and North America (Fig. 2.33).

Wind action is confined largely to arid and semi-arid areas and to coastal positions where the formation of sand dunes is a most characteristic and conspicuous process. However, in the past

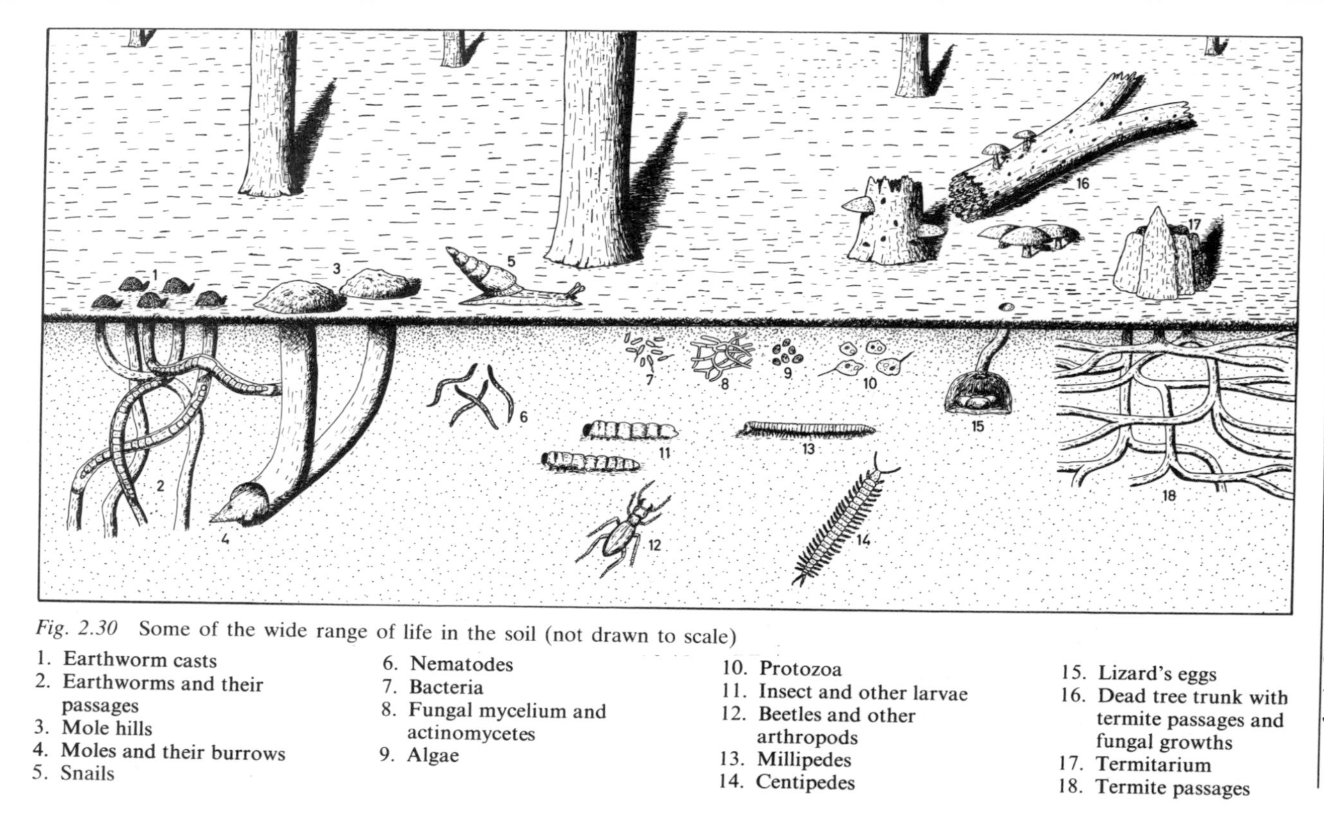

Fig. 2.30 Some of the wide range of life in the soil (not drawn to scale)

1. Earthworm casts
2. Earthworms and their passages
3. Mole hills
4. Moles and their burrows
5. Snails
6. Nematodes
7. Bacteria
8. Fungal mycelium and actinomycetes
9. Algae
10. Protozoa
11. Insect and other larvae
12. Beetles and other arthropods
13. Millipedes
14. Centipedes
15. Lizard's eggs
16. Dead tree trunk with termite passages and fungal growths
17. Termitarium
18. Termite passages

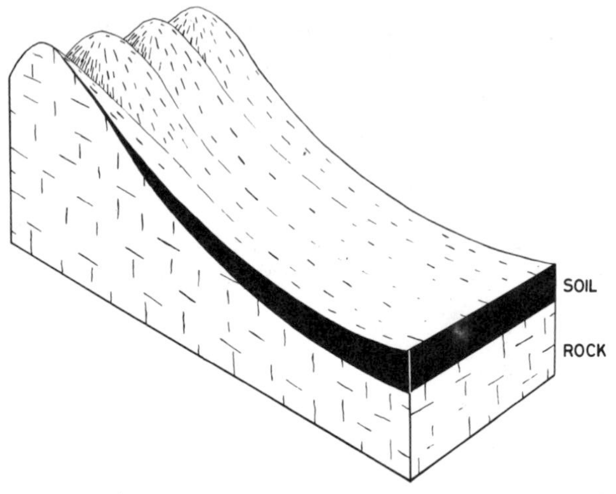

Fig. 2.31 The effect of topography on soil depth; with decreasing inclination the soil increases in thickness because the site is more stable

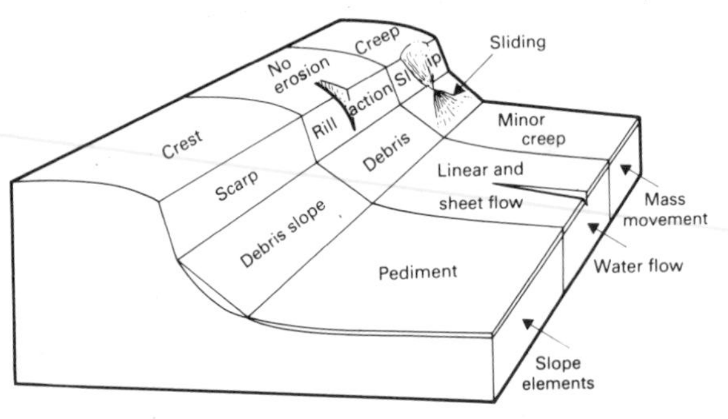

Fig. 2.32 Four element landscape model

wind activity was responsible for the silt deposits (loess) which are so common in parts of Australia, China, the USSR, the USA and elsewhere (Figs. 2.34 and 2.35).

Repeated freezing and thawing of wet unconsolidated material on slopes cause a considerable amount of movement. This process is known as solifluction and is probably the principal geomorphological process taking place in polar areas and may lead to the accumulation of thick deposits of material on the lower parts of slopes and the formation of terraces.

Fig. 2.33 Stratified glacial deposits in eastern Scotland. The two lower strata are unsorted compact till with angular stones and boulders. The upper stratum is a glacio-fluvial deposit hence the higher concentration of stones and boulders, many of which are rounded or sub-rounded

The movement *en masse* of material down slopes is probably more common than is realised. This may be gradual and may therefore pass unnoticed but when it is rapid as in the form of landslides following earthquakes or torrential rain the amount of

Fig. 2.34 Sand dunes in the Sahara desert

material that moves can be considerable and catastrophic (Fig. 2.36).

Time

Soil formation is a very long and slow process requiring thousands and even millions of years. Since this is much greater than the life span of an individual human being, it is impossible to make categorical statements about the various stages in the development of soils.

Not all soils have been developing for the same length of time but most started their development at various points during the last hundred million years.

Some horizons develop before others, especially those at the surface which may take only a few decades to form in unconsolidated deposits. Horizons requiring a considerable amount of rock weathering may take more than a million years to

Fig. 2.35 A loess deposit in central Poland. Note the characteristic prismatic structure in the upper part of the section and the dark coloured buried soil near the base

develop. Three examples of soil formation or the development of soil through time are given on page 73 *et seq*.

Fig. 2.36 A landslide. The material is a laminated lacustrine clay which slid over a highway

Soil evolution

Initially pedologists tended to interpret most soil features as the result of the interaction of the prevailing environmental conditions at the time when soil examinations were made. However, it is now evident that most places have experienced a succession of different climates which have induced changes in the vegetation and in soil formation. Therefore, most soils are not developed by a single set of processes but undergo successive waves of pedogenesis and are therefore polygenetic. Furthermore each wave implants certain features that are inherited by the succeeding phase or phases. The occurrence of strongly weathered soils in the lower Saharan and Australian deserts are clear indications of some of the climatic and soil changes that have taken place, since these soils are considered to have formed initially under hot humid conditions but now occur in deserts. Similarly, many of the soils of Europe and North America show the influence of the cold conditions that existed when glaciers extended further to the south.

In historical times many evolutionary paths have been directed by human activities; the polders of Holland and the puddled soils to grow rice in the Far East quickly spring to mind. Many plant communities are produced by human practices; excellent examples are the *Ericaceous* heaths of Europe and many of the savannas of the tropics, both of these plant communities being maintained by systematic burning.

Some workers, particularly Jenny (1941), have tried to demonstrate quite unconvincingly that these factors are independent variables, i.e. each of them can change and vary from place to place without the influence of any of the others. Only time can be regarded as an independent variable; the other four depend to a greater or lesser extent upon each other, upon the soil itself or upon some other factor. For example, it is now generally accepted that the various plant communities are largely a function of climate which is a function of wind currents, latitude, proximity to large bodies of water and elevation.

Many attempts have been made to show that some factors are more important than others and therefore play a major role in soil formation. Such efforts are a little unrealistic since each factor is absolutely essential and none can be considered as more important than any other, although locally one factor may exert a particularly strong influence.

3 *Processes in the soil system*

Soils are complex, dynamic systems, in which an almost countless number of processes are taking place. Generally these processes can be classified as chemical, physical and biological but there are no sharp divisions between these three groups. For example, oxidation and reduction are usually regarded as chemical processes but they can be accomplished by microorganisms; similarly the translocation of mineral particles can take place either in suspension or in the bodies of organisms such as earthworms.

Chemical processes

The main chemical processes include hydration, hydrolysis, solution, clay mineral formation, oxidation and reduction.

Hydration

Hydration is the process whereby substances absorb water. Few of the primary minerals undergo hydration directly, therefore very little takes place during the early stages of weathering. The principal exception is biotite which absorbs water between its layers, expands, and finally splits apart (Fig 3.1). Hydration is more often a secondary process affecting decomposition products such as iron and aluminium oxides.

Hydrolysis

Hydrolysis is probably the most important process participating in the destruction of minerals and soil formation. It is the replacement of cations such as calcium, sodium and potassium in the structure of the primary silicates by hydrogen ions from the soil solution, eventually leading to the complete decomposition of the minerals. The products of hydrolysis such as calcium

Fig. 3.1 Splitting apart of biotite following hydration

are then available to be taken up by plants or removed by water flowing through the soil or they may precipitate out of solution.

Solution

There are only a few substances found in soils that are soluble in water or carbonic acid. Nitrates, chlorides and sulphates are very soluble but these only occur in appreciable amounts in the soils of arid areas. Calcite and dolomite are less soluble but are widespread and form the major components of limestone, chalk and some other parent materials. These materials are very distinctive since they are almost completely soluble in carbonic acid and therefore supply only a very small residue after solution. Consequently soils developed on these materials are normally quite shallow. Limestone exposed at the surface usually develops very distinctive solution features known as lapies and shown in Fig 3.2. Less soluble is apatite (calcium phosphate) which can persist for thousands of years in some soils of humid areas devel-

Fig. 3.2 The characteristic solution features or lapies in limestone

oped in drift deposits. On the other hand, most other materials, particularly the silicate rocks, furnish a considerable residue of primary and secondary products.

Some minerals such as quartz which are usually considered to be inert and insoluble do dissolve eventually. This accounts for the small amount of primary material <50 μm found in many of the very old soils of humid tropical and subtropical areas.

Products of hydrolysis and solution

These include the weathering solution, the resistant residue and alteration compounds. The weathering solution contains the basic cations together with some iron, aluminium and silicate ions which are partly or completely lost, redistributed in the soil system or taken up by plants. The resistant residue includes quartz, zircon, rutile and magnetite which alter only very slowly but do decompose when present as very small particles.

The alteration compounds are principally compounds of iron and aluminium; silica and clay minerals. Iron forms ferric hydroxide, goethite – αFeO-OH or hematite – Fe_2O_3. Ferric hydroxide is an amorphous yellowish-brown substance that occurs in many soils in their initial stages of formation. Goethite is crystalline with reddish-brown colour but changes to yellowish-brown as it becomes hydrated. Goethite has a wide distribution, ranging from the tropics to the arctic and is one of the main colouring substances in soils. Hematite is bright red and occurs chiefly in soils of tropical and subtropical areas or in old geological formations. The other iron compound is lepidocrocite which is bright orange in colour, occurring principally in soils subjected to periodic waterlogging.

In a crystalline form, aluminium hydroxide occurs mainly in the soils of humid tropical and subtropical areas as gibbsite $\gamma Al(OH)_3$ which is also the principal constituent of bauxite.

Amorphous or hydrous silica, in addition to forming part of the clay, may be lost in the drainage water or redistributed within the soil system. Sometimes it may accumulate as opal and cement the soil into a massive rock-like material known in Australia as silcrete.

Finally there is manganese dioxide which is of restricted distribution being found as the blue black coating within the soil or associated with compounds of iron in certain concretionary and massive deposits.

Transformation of individual minerals

The breakdown of the individual minerals depends largely upon the nature of the climate, thus the products of decomposition of a given mineral will vary from place to place. For example, feldspars can be changed to mica in a cool climate, to kaolinite in a hot, moderately humid climate and to gibbsite in a very hot, very humid climate. This wide variability in the end products of weathering applies also to the amphiboles and pyroxenes. The weathering of minerals is often controlled by their crystal form. Figure 3.3 shows the early stage in the weathering of a feldspar. In this case there is differential weathering presumably due to small variations in composition.

Clay minerals – their formation and properties

Early workers visualised a simple transformation from primary minerals such as feldspars to clay minerals like kaolinite but it

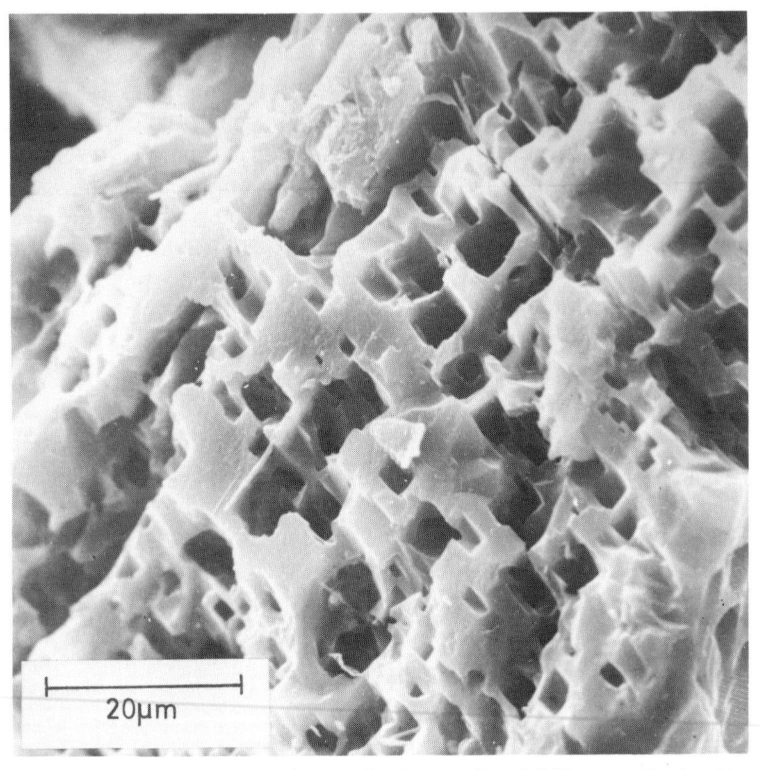

Fig. 3.3 Scanning electron micrograph of a weathered feldspar grain showing pitting and differential weathering due to small differences in composition

has been shown that the structures of these two minerals are so different that this is not possible. For many situations it seems that the primary minerals undergo fairly complete decomposition to simple substances followed by the synthesis of the individual clay minerals. This involves principally the rearrangement of the silicon-oxygen tetrahedra and the aluminium octahedra both of which become aligned to form sheets within the clay minerals.

One of the most important properties of clay minerals is that they have negative charges which allow them to adsorb and exchange cations on their surfaces i.e. they have a *cation exchange capacity* (see page 112), and because of variations in their structure they have widely differing cation exchange capacities. Also they have a capacity to absorb water. Kaolinite has a low cation exchange capacity (30–150 meq kg^{-1}) and expands very slightly when wet. On the other hand, the exchange capacities of vermiculite (1000–1500 meq kg^{-1}) and montmorillonite (800–1500 meq kg^{-1}) are high and both of these minerals can absorb water and expand, particularly montmorillonite. Hydrous mica occupies an intermediate position with a moderate cation exchange capacity (100–140 meq kg^{-1}) and a small capacity for swelling when wet. The occurrence of these minerals in soils is related largely to pH. Kaolinite forms in acid soils with mica, vermiculite and montmorillonite forming under progressively more alkaline conditions.

Flocculation and dispersion are two more properties of clays. Flocculation is the reaction whereby the individual particles of clay coagulate to form floccular aggregates. This can be demonstrated in the laboratory by adding a small amount of calcium hydroxide to a suspension of clay. Immediately floccular aggregates form and gradually settle to the bottom of the vessel – a similar reaction takes place in soils to form crumbs and granules. At the other extreme is the state of dispersion in which the individual particles are kept separate one from the other by a few ions particularly sodium. Thus depending upon the nature of the cations present in the soil it may either be in a flocculated or dispersed and often massive condition (see page 99).

Clay minerals scarcely ever occur in a pure form in soils; often a single particle is composed of interstratified layers of two or more different clay minerals, or they may be enmeshed in large quantities of other secondary materials hence the many difficulties encountered in their identification.

Oxidation and reduction

It is convenient to consider these two processes together since one is the reverse of the other. Iron is the principal substance affected by these processes for it is one of the few elements that is usually in the reduced state in the primary minerals. Consequently when it is released by hydrolysis and enters an aerobic atmosphere it is quickly oxidised to the ferric state and precipitates as ferric hydroxide to give yellow or brown colours. If, on the other hand, the iron is released into an anaerobic environment it stays in the ferrous state. Such soils range in colour from blue to grey to olive to black depending upon the precise compound that is formed. Vivianite (ferrous phosphate) imparts blue colours while black colours are due to sulphides which often form in coastal and estuarine marshes. The formation of horizons by partial or complete reduction is often referred to as gleying. There is also biological oxidation and reduction. The ochre sometimes deposited in drains and in shallow slow flowing streams is a microbiological oxidation product of iron.

Under anaerobic conditions nitrogen in the form of ammonia or nitrate is reduced to N_2O or N_2 and lost into the atmosphere. This can be very significant in some soils leading to the inefficient use of fertilizers.

Physical processes

The main physical processes are translocation, aggregation, freezing and thawing, and expansion and contraction, but the agencies responsible are very varied.

Aggregation

Aggregation is the process whereby a number of particles are held or bound together to form units of varying but characteristic shapes. This property is discussed on page 94.

Translocation

Many of the processes of soil formation and horizon differentiation are concerned primarily with removal, reorganisation and redistribution of material in the upper 2 m or so of the Earth's crust.

Percolating through the soil in a humid environment is a large volume of water which on moving downwards takes with it dissolved material some of which may be translocated to a horizon below or it may be lost in the drainage water. On slopes, some moisture moves laterally causing the soils at lower positions to be enriched by soluble substances. Sometimes, the whole soil may become saturated by water moving laterally and there is free water at the surface. Such conditions are known as *flushes* and usually carry a plant community indicative of a moist habitat.

Fine particles and colloidal materials are often transported in suspension from one place to another within the soil system. Perhaps the most important manifestation of this process is the removal of particles $<0.5\ \mu\text{m}$ from the upper horizons of some soils followed by their deposition in the middle position to form clay coatings (see page 105). Ultimately this can lead to differences of more than 20 per cent in the clay content between adjacent horizons.

As stated earlier many members of the mesofauna and some small mammals are responsible for redistributing large amounts of material within the soil.

Freezing and thawing

These two processes take place to varying degrees over a wide area of soils. During freezing a number of ice patterns develop as determined by a number of factors. Needle ice usually forms near to the surface of the ground and is responsible for heaving stones to the surface and the breakdown of large clods, hence the reason for ploughing before the onset of winter. Below the surface, freezing causes the formation of ice veins surrounding lenses of soil and in polar areas repeated freezing and thawing cause profound disturbance of the soil and the development of a number of characteristic patterns such as stone polygons and mud polygons (Figs. 3.4 and 3.5).

Freezing and thawing also cause rocks to be shattered leading to the development of extensive areas of angular rock fragments and as already mentioned on page 50 a considerable amount of solifluction takes place in polar areas as a result. These processes also cause the disruption of the foundations of buildings and the misalignment of rail tracks.

Expansion and contraction

These processes take place mainly as a result of wetting and drying and are very important in soils containing a high proportion of clay with an expanding lattice such as montmorillonite. They occur principally in the soils of hot environments with alternating wet and dry seasons. Contraction leads to the formation of wide and deep cracks causing the roots of the vegetation to

Fig. 3.4 Stone polygons 2–3 m wide, on Baffin Island

be stretched and broken. During expansion high pressures are developed within these soils, causing rupture and the slippage of one block over the other and the formation of polished faces or *slickensides* on the blocks. Expansion and contraction also cause the formation of micro-topographic features known as *gilgai* (Figs. 3.6 and 3.7). They can also cause the disruption of the foundations of buildings.

Biological processes

The most important biological processes taking place in soils are the decomposition of organic matter or humification, nitrogen transformations and the translocation of material from one place to another. At the start of soil formation the contribution of the dead bodies of microorganisms, particularly algae, can be very

Fig. 3.5 A mud polygon 120 cm wide from northern Alaska

important in the initiation of a number of cycles, particularly the nitrogen cycle.

Humification

The decomposition of organic matter or humification is an extremely complex process involving various organisms including fungi, bacteria, actinomycetes, worms and termites. It has been shown that in the case of pine needles the process is very slow with invasion and decomposition by the fungus *Lophodermium pinastri* starting when the needles are still on the tree and continuing after they have fallen onto the ground. This fungus causes the formation on the needles of characteristic black spots and transverse black bars which are clearly seen in hand specimens and thin sections (Figs. 3.8 and 3.9). When the needles are on the ground they are attacked by wave after wave of fungi, each wave being fairly specific in its action and each successive wave being capable of breaking down more complex plant compounds than the previous wave (Fig. 2.26; p. 43). At first

Fig. 3.6 Circular raised areas of gilgai

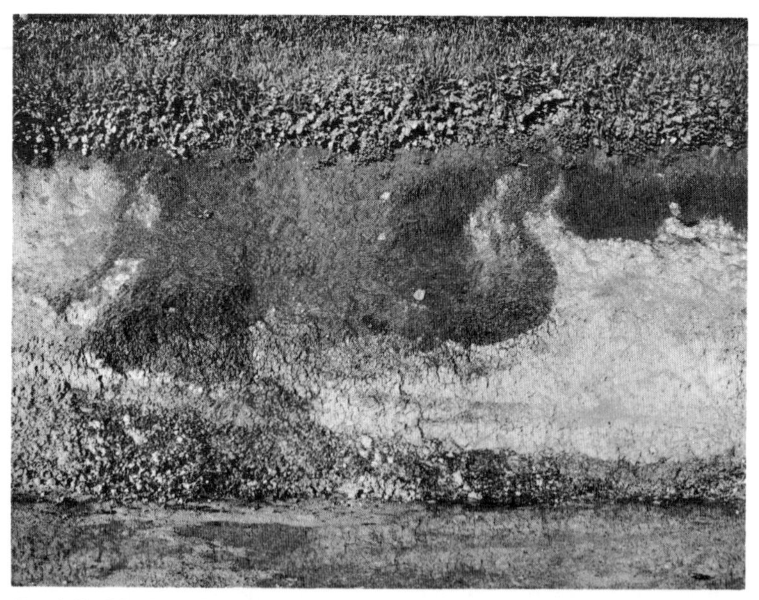

Fig. 3.7 Gilgai profile. The pressures produced by expansion and contraction cause contortions in the soil and heaving of the underlying material to the surface

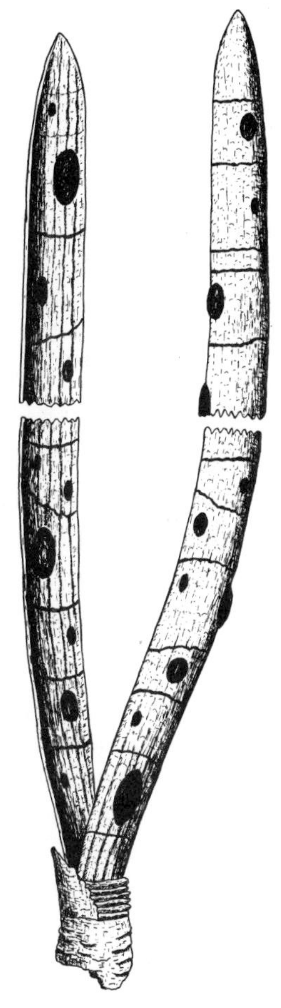

Fig. 3.8 Pine needle infected with the fungus *Lophodermium pinastri* as indicated by the characteristic black spots and bars

the simple compounds such as starches and sugars are decomposed, followed by the proteins, cellulose, hemicellulose and finally the very resistant compounds such as tannins. Gradually the material is decomposed and the dark coloured substance known as *humus* is formed, the whole process taking about seven to ten years.

Small arthropods such as mites also play a part in the process by eating the softer inner parts of the needles as well as

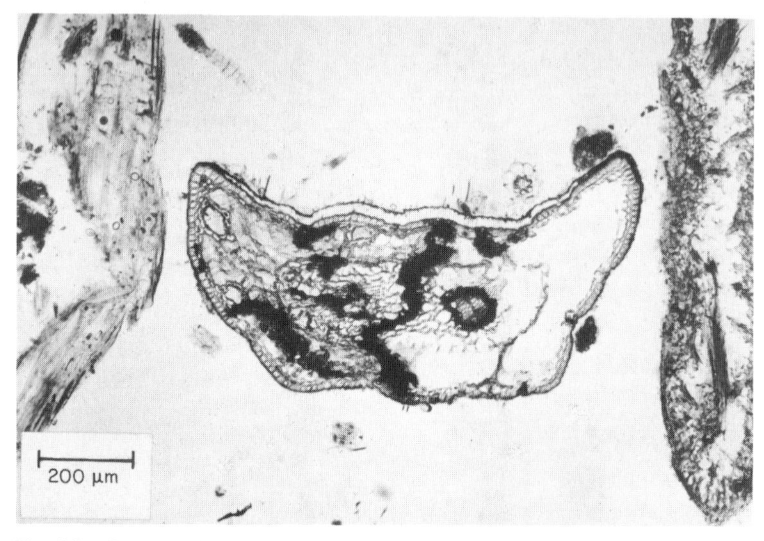

Fig. 3.9 Cross-section through a pine needle showing the early stages of decomposition by the fungus *Lophodermium pinastri*. The fungus has produced the black material within the needle

consuming some of the fungal mycelium (Fig. 2.27; p. 43). For these reasons the organic matter beneath pine forests shows progressive decomposition with depth. In contrast the decomposition of the leaves of deciduous trees by earthworms and bacteria is usually much more rapid and may be accomplished within a year, leaving little organic matter at the surface.

Generally the formation, decompostion and redistribution of organic matter can be represented by the carbon cycle in Fig. 3.10.

Nitrogen transformations

The main forms of nitrogen transformation are *ammonification*, *nitrification* and *nitrogen fixation*. Ammonification is the process whereby nitrogenous compounds in plant and animal tissues are decomposed to produce ammonia which is changed by nitrification into nitrite, and then into nitrate, each stage being accomplished by specific microorganisms. The formation of ammonia is accomplished by heterotrophic bacteria but the two other stages are brought about by autotrophic bacteria. Ammonia is oxidised by *Nitrobacter, Nitrosomonas* and *Nitrococcus* and the nitrite is oxidised by *Nitrobacter*. These processes require aerobic conditions; if the soil is waterlogged for any length of time the nitrogenous compounds are reduced by denitrification to nitrogen which is lost to the atmosphere.

Nitrogen fixation is the process during which soil bacteria take up nitrogen from the soil atmosphere to form their body protein. The organisms include *Azotobacter*, *Clostridium pasteurianium* and *Beijerinckia* which upon death enter the nitrogen cycle and are decomposed to form nitrate for plant uptake.

There are also a number of bacteria which enter the roots of certain plants, particularly members of the *Leguminoseae*. There they multiply, form nodules and fix atmospheric nitrogen which then passes into the conducting system of the plant as an essential element (see page 45).

Since nitrogen does not form part of the lithosphere all the nitrogen that forms part of the tissues of plants and animals must come initially from the atmosphere, being transformed in the soil by various microorganisms. Therefore, in one sense, nitrogen fixing microorganisms can be regarded as the corner stone of **all life**.

The above processes may appear to be of little significance in the direct formation of soils but they are of paramount importance in plant nutrition. Each of the above processes can be regarded as part of a cycle during which nitrogen ions pass from one stage to another as shown in Fig. 3.11.

Translocation

Some emphasis must be given to those biological processes that bring about churning and translocation. The most dramatic manifestations of these processes are brought about by soil-inhabiting vertebrates but probably the greatest amount is accomplished by earthworms, enchytraeid worms and termites (Figs. 2.23, 2.28, 2.29 and 2.30).

In conclusion it must be stressed that soil-inhabiting fauna and flora contribute enormously to the biological balance in nature for without them the remains of plants and animals would accumulate to very great thicknesses. This is clearly demonstrated by the formation of peat which is the accumulation of

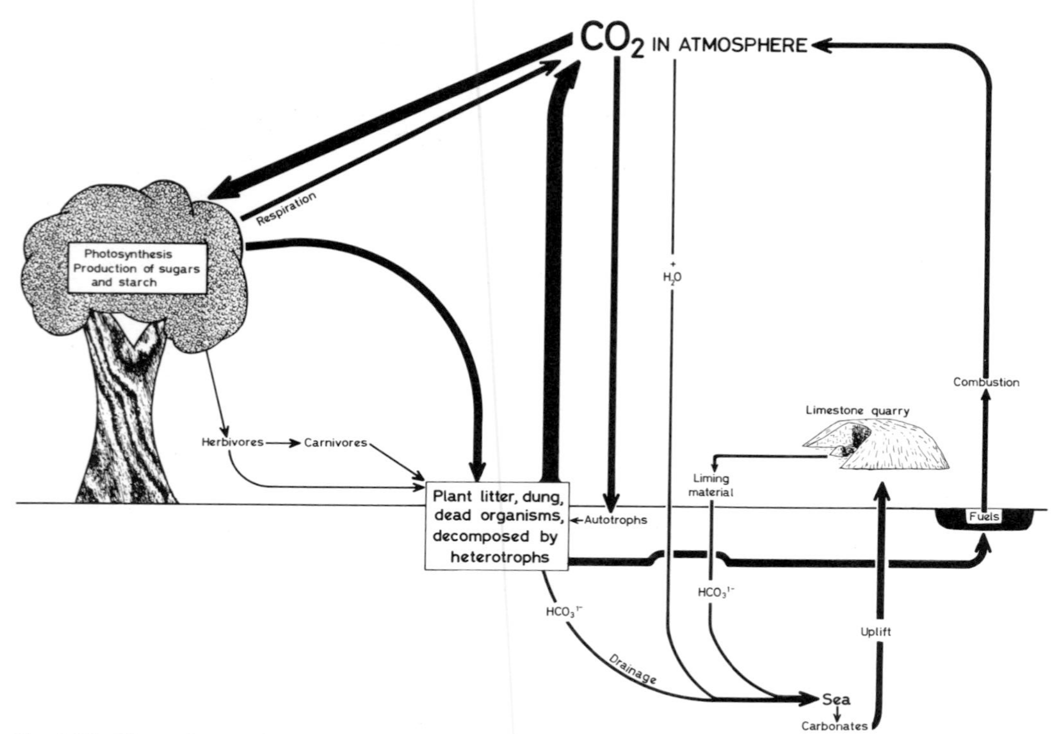

Fig. 3.10 The carbon cycle

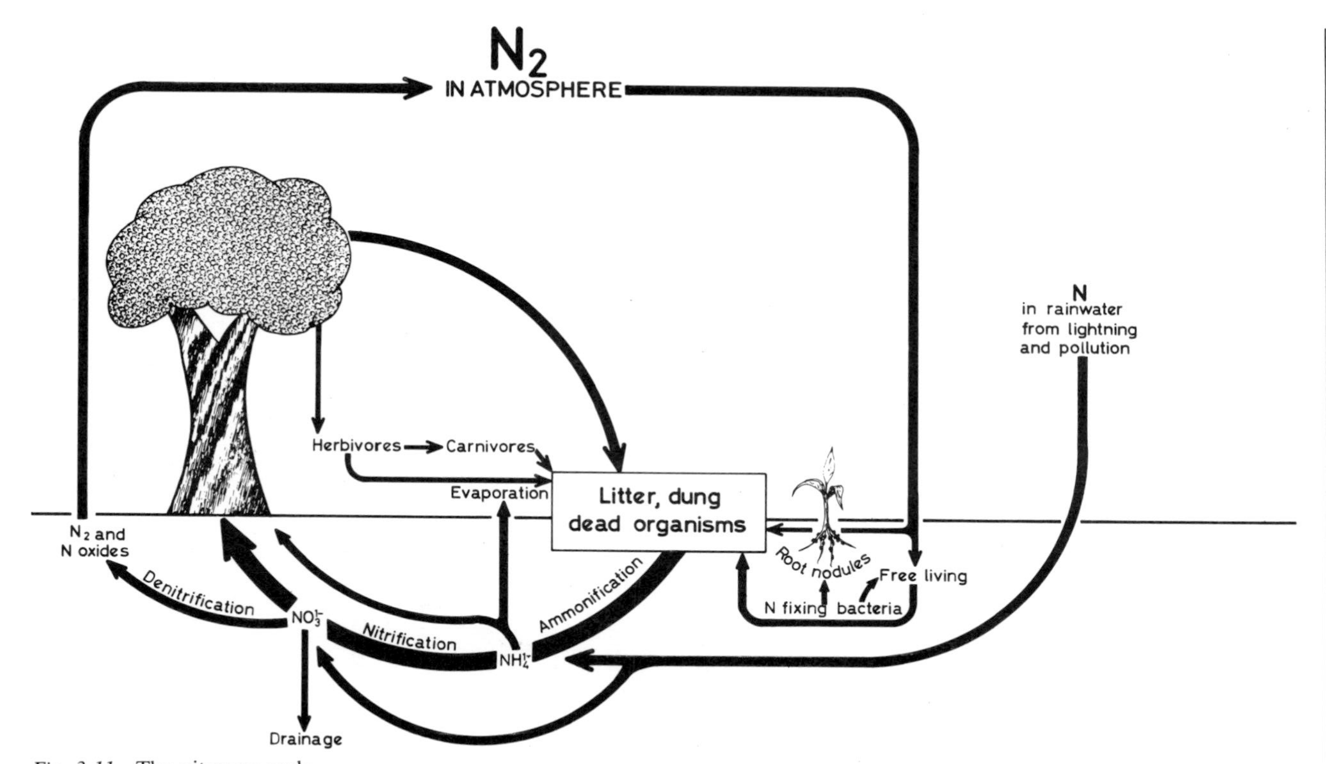

Fig. 3.11 The nitrogen cycle

organic matter under wet anaerobic conditions which are not suitable for litter decomposing organisms; hence its accumulation (Fig. 8.5).

Among the organisms in the soil there exists an extremely complex inter-relationship for seldom does a single type of organism exist or operate separately from the others. Some highly contrasting organisms coexist while others are predators, competitors or parasites. Earthworms and bacteria coexist, for when earthworms ingest a mixture of organic and mineral material they also take in large numbers of microorganisms, particularly bacteria, because the alimentary systems of earthworms do not seem to have a distinctive bacterial flora but is the same as that of the surrounding soil. These bacteria, together with the enzymes liberated in the alimentary systems of the worms are responsible for the breakdown of the organic matter, thus supplying energy and body tissue for the worms. Eventually the worms die and are themselves decomposed by other microorganisms or they may be eaten by a mole or other predators. Some microorganisms produce vitamins that are utilised by other microorganisms.

Another example is for plant tissues to be softened by fungal growth and then eaten by small arthropods which at the same time may eat some of the fungus. Later when the arthropod dies it will be decomposed by other microorganisms. Some nematodes live on protoplasm from living cells but they can be trapped and killed by a special loop mechanism of certain fungi and then decomposed by the fungus to form its food.

There are also the protozoa which seem in part to be scavengers by devouring small fragments of organic matter, but they also consume large amounts of bacteria and may control their numbers.

A fascinating example of the indirect effect of microorganisms is found in Western Australia. There, the soils and weathered rocks are many tens of metres thick and over a period of many millennia they have accumulated soluble salts, particularly sodium chloride. The salts are not harmful when the native Jarrah forest remains undisturbed but large areas are being attacked and destroyed by the root invading fungus *Phytophthora cinnamomi*. This allows more moisture to enter and percolate through the soils thereby leaching the salts from the high to the low ground and into streams and rivers. Thus the forest is being destroyed, the soils in the lower parts of the landscape are becoming saline and unsuitable for most plants, and domestic water supplies rendered useless. This devastation by a fungus should be contrasted with the nitrogen fixing bacteria which are the cornerstone of life. (Removal of the forests by humans has the same effect.)

Thus it is seen that there are a number of different but interrelated facets to life in the soil.

Soil formation

The above factors and processes combine to form the very wide range of soils that occurs on the surface of the Earth, and is discussed in Chapter 7. The general trends are summarised and presented diagrammatically in Figs. 3.12 and 3.13. In addition the formation of a Podzol from a sand dune, a Ferralsol from rock and a Chernozem from loess will be considered. These three specific examples illustrate some of the more important principles in soil formation. Podzols form in cool humid conditions in which translocation and accumulation of sesquioxides in the middle horizon is the main process. In the Ferralsols progressive weathering under hot humid conditions predominates, while Chernozems form in cool semi-arid areas where there is vigorous faunal activity and translocation of carbonates.

Stages in the formation of a Podzol – Fig. 3.14

Initially there is the Regosol stage in which there is a thin litter at the surface overlying a crude mixture of organic and mineral material about 10–15 cm thick. Below this is the unaltered sand. This stage is of short duration and is quickly followed by the accumulation of a thick litter with progressively more decomposed organic matter with depth. Below, there is a dark grey homogeneous mixture of organic and mineral material up to 15 cm thick and followed by some rusty staining of the sand. This staining is due to the deposition of iron and marks the initial stage in the formation of the middle horizon. This is then a Ranker with only a well developed upper horizon. The next stage is the fully developed Podzol taking less than a thousand years

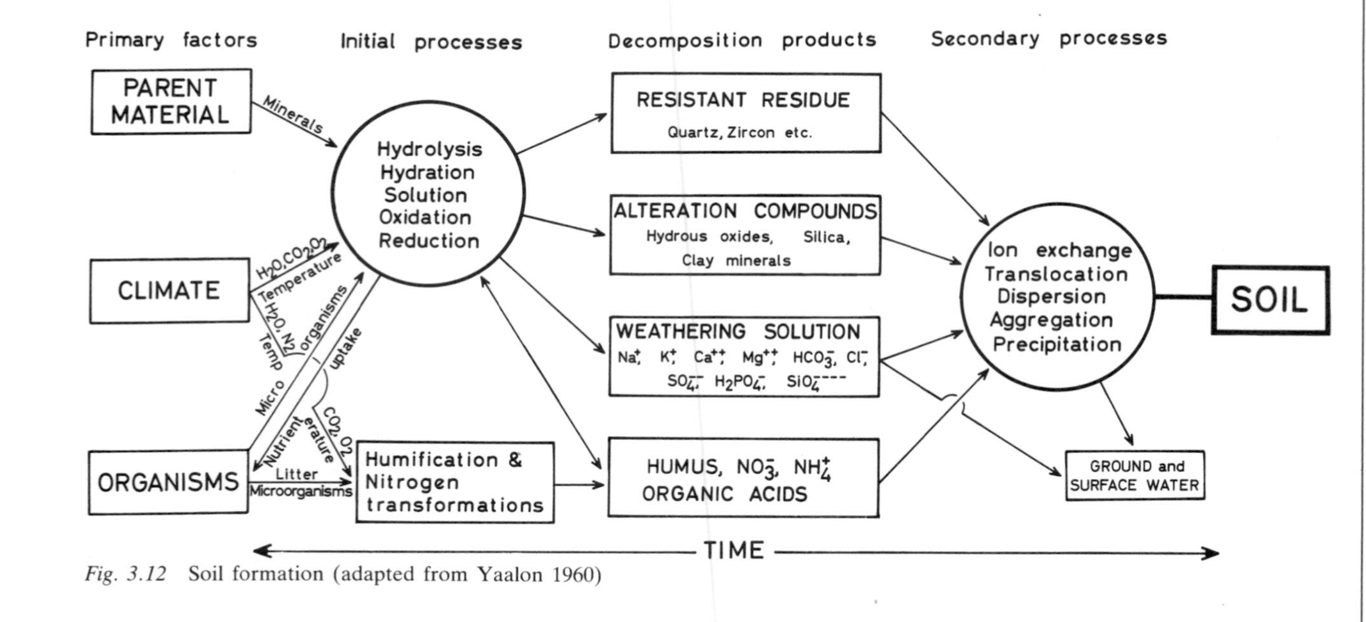

Fig. 3.12 Soil formation (adapted from Yaalon 1960)

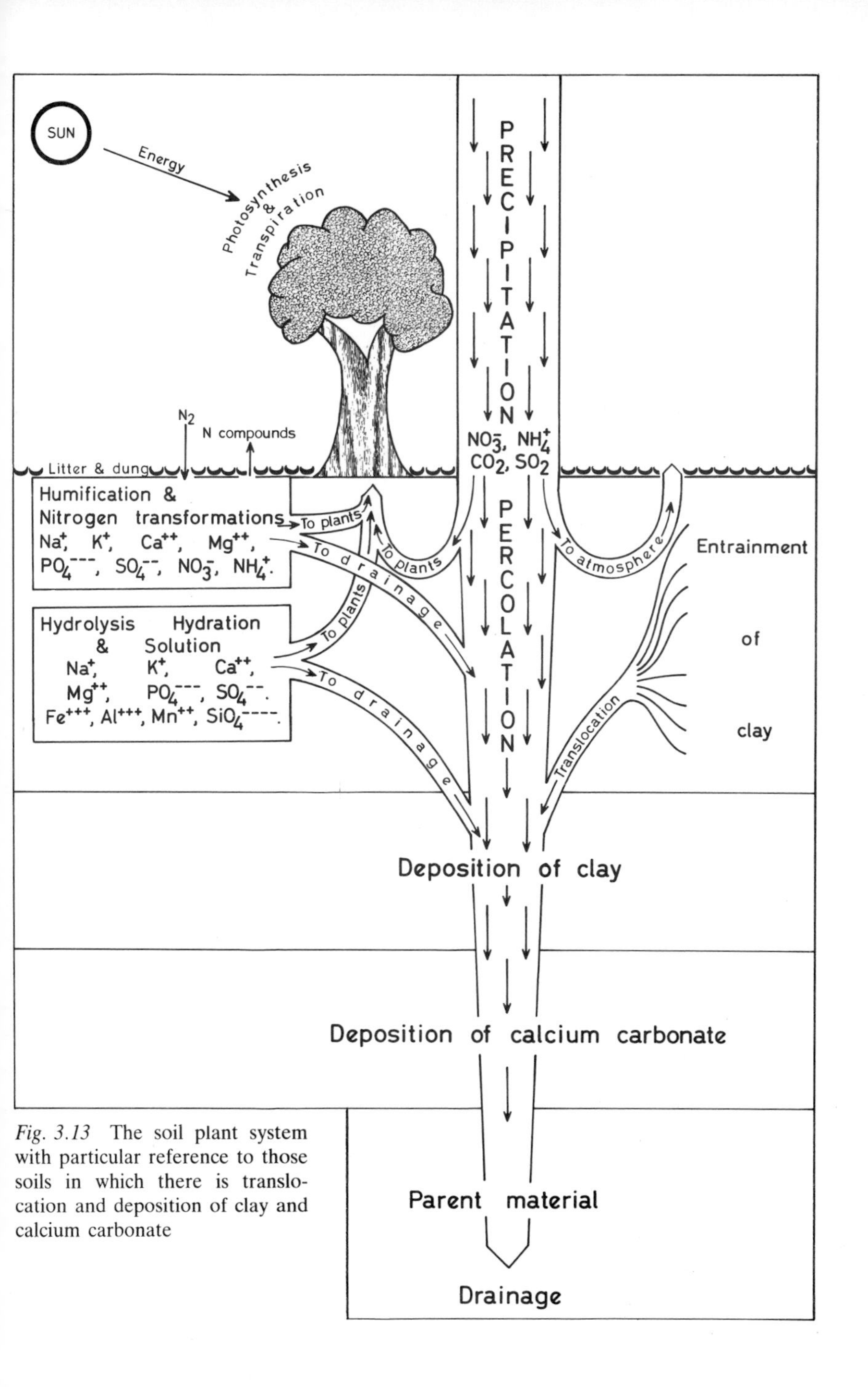

Fig. 3.13 The soil plant system with particular reference to those soils in which there is translocation and deposition of clay and calcium carbonate

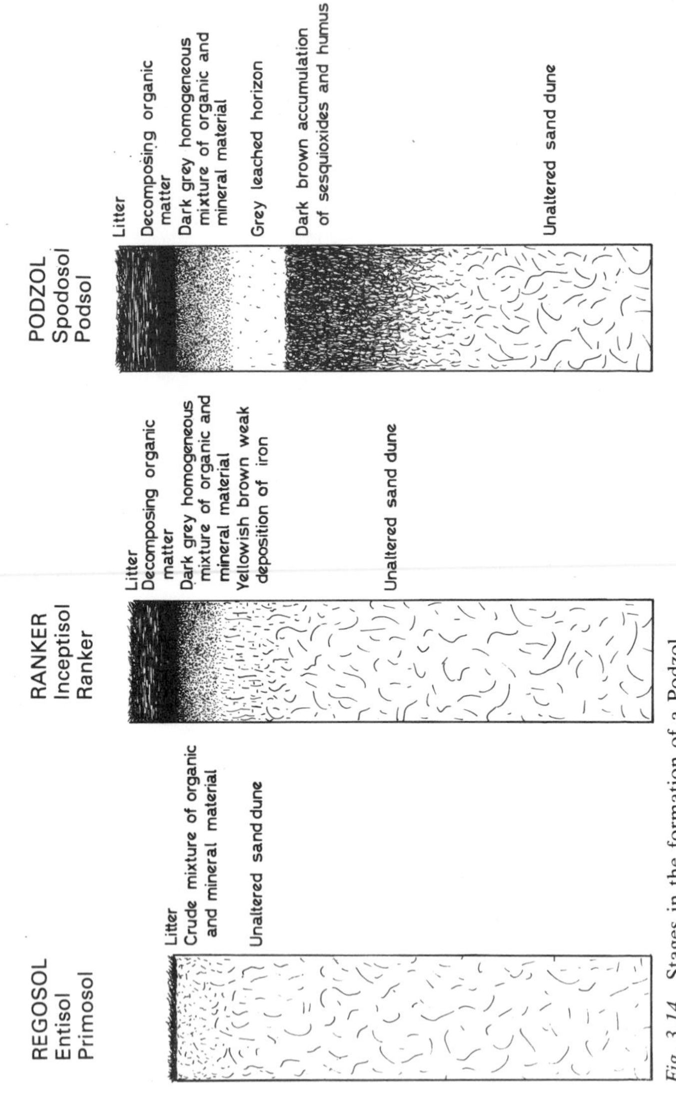

Fig. 3.14 Stages in the formation of a Podzol

to form and characterised by the development of a marked bleached horizon due to the leaching out of iron, aluminium and humus followed by their deposition below. In some cases sufficient material accumulates to cause cementation and hardening of this horizon. These three stages may take place at the same site or they may be found in progressively older sand dunes. Accompanying the development of the soil there is usually a plant succession starting with mosses, lichens, grasses and herbs and culminating in pine forest.

Stages in the formation of a Ferralsol – Fig. 3.15

At stage 1 there is the fresh rock surface followed by the formation of Rankers in small depressions where material can accumulate. Gradually the rock weathers during which some of the products remain while others are lost in solution from the system, hence the lowering of the surface. This is the Cambisol stage at which the soils are still fairly shallow and contain a high content of primary minerals. With time the soil deepens, and all of the primary minerals such as feldspars, amphiboles and pyroxenes are decomposed and the iron is oxidised to give the characteristic red colour of the Ferralsol at stage 4. In contrast to Podzols, Ferralsols develop very slowly because of the large amount of hydrolysis involved and may take over one hundred thousand years to form. There is also a plant succession on these soils with tropical rain forest as the final stage.

Stages in the formation of a Chernozem – Fig. 3.16

Probably the greatest areas of Chernozems are developed in loess which is a fine wind-blown sediment deposited during the last glaciation. Following the deposition of loess there is colonisation of the site by vegetation and the formation of a thin crude mixture of organic matter and mineral soil to give a Regosol. This is followed by a denser growth of grass and the formation of a Ranker in which there is a fairly homogeneous blend of organic and mineral material resulting from a vigorous earthworm and blind mole rat population. There is also a small amount of translocation of calcium carbonate. Ultimately a deep

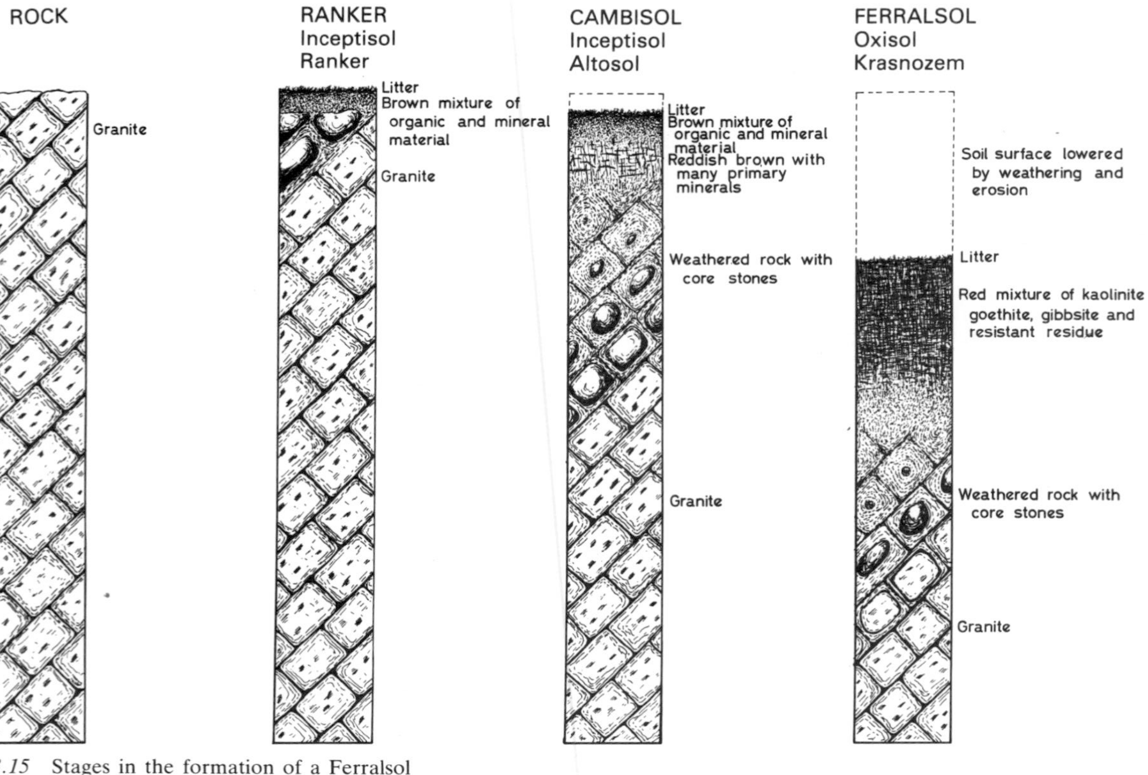

Fig. 3.15 Stages in the formation of a Ferralsol

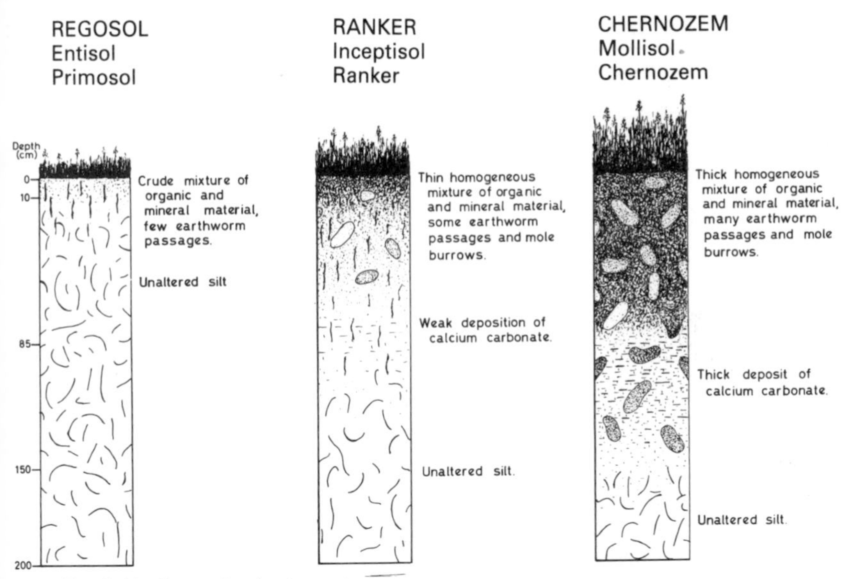

Fig. 3.16 Stages in the formation of a Chernozem

Chernozem forms mainly through the continued activity of the soil fauna incorporating organic matter and mild leaching causing the translocation of calcium carbonate. Like Podzols, Chernozems will form within a thousand years.

4 *Properties of soils*

Because soils develop as a result of the interaction of the factors of soil formation they have a wide range of properties. The most common are listed and discussed below. They include those properties of importance in soil formation and soil classification. Properties of importance in agriculture are given in Chapter 6 while those important to engineers are given at the end of this chapter.

1. Surface of the soil
2. Position of horizons
3. Thickness of horizons
4. Boundaries of horizons
5. Consistence – handling properties
6. Mineral composition
   - Larger separates – material >2 mm
   - The fine earth – material <2 mm
   - Texture
7. Colour
8. Fabric of soil materials
9. Fine material and matrix
10. Structure and porosity
11. Atmosphere
12. Passages – faunal and root
13. Faecal material
14. Coatings
15. Moisture status
16. Density
17. pH
18. Organic matter
19. Cation exchange properties
20. Soluble salts
21. Carbonates
22. Elemental composition

23. Amorphous oxides – free silica alumina and iron oxides
24. Concretions
25. Engineering soil properties

1. Surface of the soil

The soil surface may have distinctive features such as mole hills (see page 40), frost features (see page 64) or gilgai (see page 66). There may also be the accumulation of stones, the formation of a crust or the development of cracks due to desiccation.

2. Position of horizons

The soil profile may be regarded as containing three positions, upper, middle and lower. The upper position being at or near to the surface usually contains horizons with the greatest amount of organic matter and strongly influenced by biological processes. The middle position often contains horizons with material washed in from above while the lower position can have a variety of materials. There may be unaltered material, a hard pan or deposits of substances such as calcite and gypsum.

3. Thickness of horizons

Most well developed horizons have fairly well defined limits of thickness. Some, such as thin iron pans are less than 1 cm whereas some others in the deeply weathered soils of the tropics may be several metres thick. The horizons in any one given soil are seldom uniform in thickness and in extreme cases form tongues into the underlying horizon.

4. Boundaries of horizons

The change from one horizon to another varies in distinctness and outline, and is usually caused by differences in colour. There are five classes of distinctness and five classes of outline as given below.

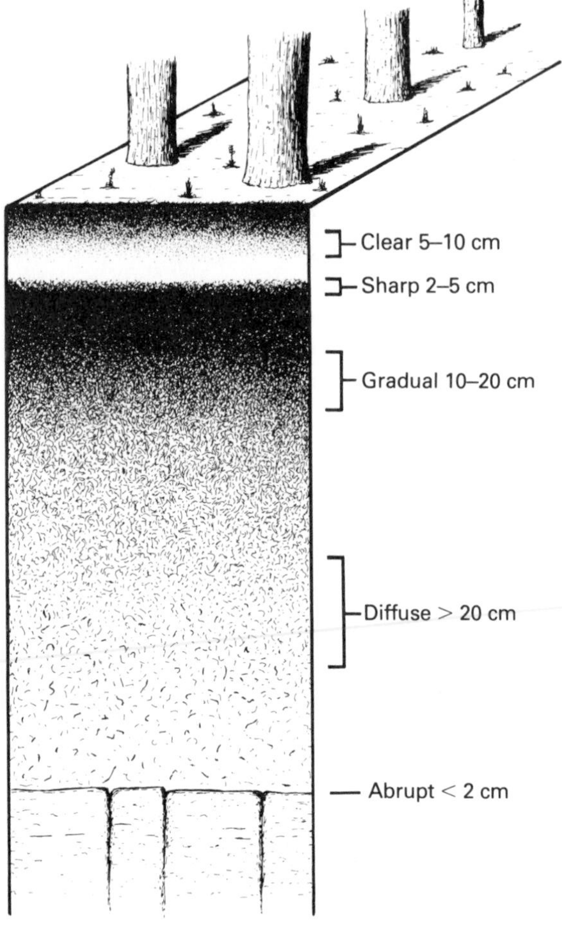

Fig. 4.1 Distinctness of soil horizon boundaries

Classes of distinctness – Fig. 4.1

Clear – change takes place within 5–10 cm
Sharp – change takes place within 2–5 cm
Gradual – change takes place within 10–20 cm
Diffuse – change takes place within >20 cm
Abrupt – change takes place within 2 cm

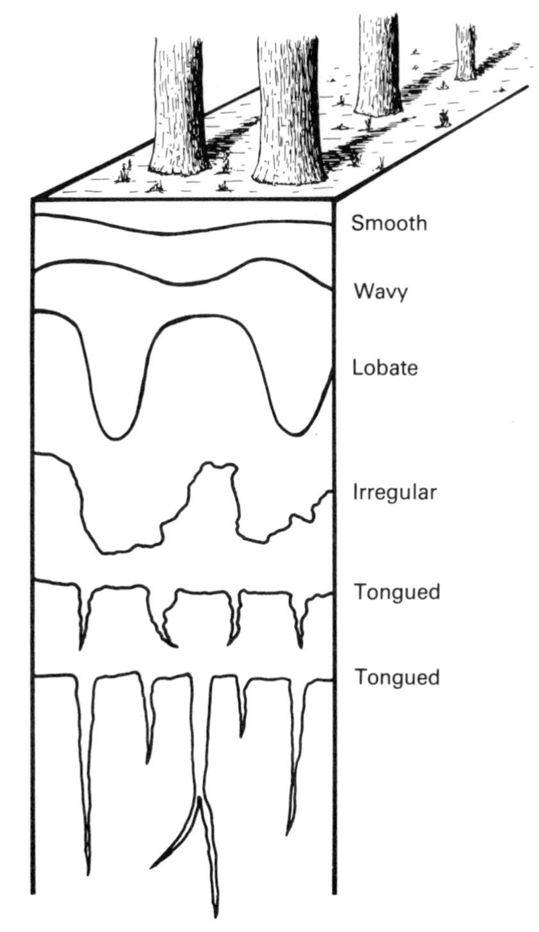

Fig. 4.2 Outline of soil horizon boundaries

Classes of outline – Fig. 4.2

Smooth – almost straight
Wavy – gently undulating
Lobate – with regular lobes
Irregular – Strongly undulating and mamillated
Tongued – forming tongues into the underlying horizon.

5. Consistence – handling properties

When soils are manipulated between the fingers and thumb they exert varying degrees of resistance to disruption and deformation as determined by their mechanical composition, degree of aggregation, content of organic matter and moisture content (see page 88). Generally the pressure needed to disrupt a dry soil increases with the content of fine material, sands are usually quite loose while clays form very hard aggregates. Moist sands have a small measure of coherence whereas moist clays are plastic and become very sticky when wet, particularly if the content of montmorillonite is high. The presence of large amounts of humified organic matter in the soil is particularly important for it increases the plasticity of sandy soils but has the reverse effect on a clay by reducing the stickiness. The consistence of soils of medium texture does not change very much with variations in moisture content. In either the dry or moist state they are usually friable, i.e. firm with well formed aggregates that crumble easily when pressure is exerted on them. When wet, these soils tend to be slightly sticky but never to the same extent as clays.

Some horizons that are massive and hard offer a considerable degree of resistance to disruption as a result of cementation by substances such as iron oxides, aluminium oxide and calcium carbonate. Resistance to disruption can result also from physical compaction.

The consistency of a soil is a very important agricultural property and it is essential for the soil to have the correct consistence at the time of cultivation. If it is too dry and hard undue strain will be placed on the implements, on the other hand if it is too wet and sticky the implements may stick and the soil may become puddled thus producing a poor seed bed for crops.

The class of consistence are:

Brittle: firm and ruptures suddenly in an explosive manner.

Compact: firm with close packing.

Firm: moderately coherent but can be crushed and broken into fragments between forefinger and thumb.

Fluffy: friable with a low bulk density.

Fluid: some soils when wet will behave like a liquid and will flow.

Friable: weakly coherent and easily crushed with gentle pressure.

Hard: dug out in coherent lumps and needs force to cause it to rupture.

Loose: non-coherent and present as single grains or aggregates.

Plastic: the moist soil can be moulded or pressed into specific shapes.

Soapy: sticky and plastic with a distinctly soapy feeling, this is characteristic of many soils having a high content of exchangeable sodium.

Soft: very weakly coherent, becoming loose with very gentle pressure.

Sticky: the soil adheres or sticks to other objects.

Tenacious: plastic but requires a considerable amount of pressure to be moulded into specific shapes.

Thixotropic: friable or firm and moist but becomes very moist or wet when continuously manipulated between the fingers.

The consistence of most materials alters with a change in moisture status. For example, some clays may be hard when dry, plastic when moist and sticky when wet.

6. Mineral composition

Soil particles are divided initially into two size classes with the limit normally set at 2 mm to delimit the "fine earth" from the larger separates including gravel.

Larger separates – material > 2 mm

A knowledge of the nature and properties of the coarse particles can often lead to important conclusions about the origin and formation of the parent material and about the soil itself. In some superficial deposits the sand and silt fractions have a similar mineralogical composition to that in the material > 2 mm. Therefore the larger separates can be used to assess the chemical composition of some soils and parent materials. This is a common field practice in many glaciated areas.

The shape of the stones often gives a clear indication of the processes which have influenced the formation of the parent material and/or the soil itself. Rounded stones occur in alluvium and beach deposits, subangular to subrounded stones in glacial drift, angular stones result from exfoliation or frost action.

Stones in superficial deposits are orientated in a number of specific patterns. The stones in alluvium usually have an imbricate pattern with their long axes aligned in the direction of river flow and dipping upstream. Glacial deposits often have stones aligned in the directions of ice flow. In areas of vigorous frost action the stones become orientated by frost heaving to form a number of patterns at the surface as well as in the soil itself (Fig. 3.4). On a flat site the stones within the soil are vertically orientated whereas on a sloping situation they become orientated parallel to the slope and normal to the contour. A similar type of orientation on slopes is associated with certain forms of soil creep found in many tropical areas.

When wind is an agent of erosion and deposition, stones at the surface become faceted by the fine particles blown against them. Such stones are known as *ventifacts*.

Rock fragments often form nuclei for the formation of other features. In horizons with calcium carbonate accumulation it is common to find some rock fragment coated with calcite or having calcite pendants, Fig. 4.3.

Stones modify the soil in many ways, they reduce the water holding capacity but cause the soil to heat up more quickly. A great disadvantage is that they increase the wear on implements and damage the crop at harvest time.

The fine earth – material < 2 mm

The material < 2 mm is divided into sand, silt and clay, the size limits of which vary between workers but normally one of the three schemes given in Table 4.1 is used.

According to the relative amounts of sand, silt and clay twelve classes have been created and presented in the form of a triangular diagram by using only three size limits as shown in Fig. 4.4.

The mechanical composition of soil horizons is the result of fairly specific processes. Soils developed in fairly recent deposits have inherited a considerable proportion of their characteristics from the parent materials but as soils increase in age, more and more clay is formed and they gradually become increasingly fine.

Fig. 4.3 Calcite pendant beneath a stone

However, the translocation of clay particles from one horizon to another may cause the horizon losing the clay to become coarser and the receiving horizon to become finer.

The shape of the particles in the sand fraction is often a useful guide to the origin of the material. Sand varies in shape from smooth and round to very rough and angular. The former is found in wind blown material and beach sand whereas rough angular sand occurs in glacial deposits. The fine earth contains both primary and secondary minerals. The former usually dominate the sand fraction and it has been found that the mineralogy of the fine sand fraction gives a fairly accurate picture of the nutrient reserve in the soil.

The clay fraction is composed predominantly of crystalline clay minerals and amorphous material. The study of this fraction yields important data, allowing important conclusions to be made about the processes taking place in the soil. The accurate determination of the particle size distribution in soils is a fairly long and tedious process but a simple approximate estimate can be carried out by determining the texture as given below.

Table 4.1 Size limits of soil separates

U.S. Department of Agriculture scheme		*International scheme*		*Massachusetts Institute of Technology*	
Name of separate	*Diameter (range)*	*Separate*	*Diameter (range)*	*Separate*	*Diameter (range)*
	Millimetres		Millimetres		Millimetres
Gravel	>2.0		Gravel >2.0		
Very coarse sand	2.0–1.0			Stones	>2.0
Coarse sand	1.0–0.5	Coarse sand	2.0–0.2	Coarse sand	2.0–0.6
Medium sand	0.5–0.25			Medium sand	0.6–0.2
Fine sand	0.25–0.10	Fine sand	0.20–0.02	Fine sand	0.2–0.06
Very fine sand	0.10–0.05				
Silt	0.05–0.002	Silt	0.02–0.002	Silt	0.06–0.002
Clay	<0.002	Clay	<0.002	Clay	<0.002

This is probably the most permanent property of soils and virtually impossible to be modified; therefore all users of soils have to devise systems of land use based on the prevailing particle size distribution. In some countries there has been continual removal of the largest stones which has led to reduced damage to implements.

Texture

The texture of soil refers to the "feel" of the moist soil resulting from the mixture of the constituent mineral particles and organic matter. Therefore it is an approximate measure of the particle size distribution or mechanical composition which is measured in

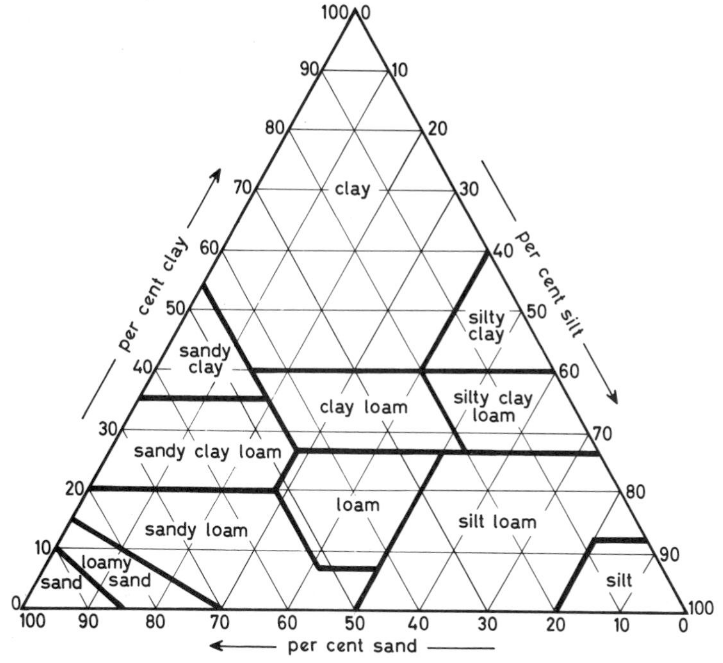

Fig. 4.4 Triangular diagram relating particle size distribution to texture according to the USDA.

the laboratory. Whereas particle size data are required for certain studies, texture as determined by feel is often more closely related to the behaviour of the soil in the field and to the physical properties of significance in agriculture.

Texture is determined by rubbing the moist soil between the fingers and thumb and is commonly determined in the field. It is a subjective technique but can be mastered with some experience and being a simple technique it has an advantage over particle size determination, which is a long tedious process.

Coarse sand particles can clearly be seen with the naked eye and have a marked gritty feel. Fine sand particles are less obvious but also have a slight to moderately gritty feel. Silt particles cannot be individually detected with the naked eye but can be seen with a 10 x lens. They have a distinctive smooth silky feel when both wet and dry but are only very slightly plastic. Clay soils are distinctly sticky and plastic when wet, but are usually compact and hard when dry.

Although different soils may have the same texture they may not have the same particle size distribution and *vice versa*. This is due mainly to variations in the amounts of organic matter, type of clay, shape of particles and degree of aggregation. It is usual for the upper horizons to contain varying amounts of organic matter. When the content is small the effect is minimal but large amounts of organic matter cause the soil to be smooth and to appear to have a higher silt content.

The type of clay will also have an effect since some clays (montmorillonite) absorb more water than others (kaolinite); thus two soils may contain the same amount of clay but the one with montmorillonite will be more sticky and plastic than that with kaolinite.

The shape of the particles can also be important. For example, when the sand grains are round the grittiness will appear to be less than if they are angular.

In some soils the clay particles are cemented to form small aggregates which cause the soil to appear to have a higher silt content. Thus a soil which has the particle size of a clay has the feel of a silty clay.

The properties of the texture classes are as follows:

Sand: extremely gritty, not smooth, not sticky or plastic, non-cohesive balls which collapse easily.

Loamy sand: extremely gritty, not smooth, not sticky or plastic, slightly cohesive balls but does not form threads.

Sandy loam: very gritty, not sticky or plastic, slightly cohesive balls, does not form threads.

Loam: moderately gritty, slightly smooth, slightly sticky and plastic, moderately cohesive balls, forms threads with great difficulty.

Sandy clay loam: moderately gritty, not smooth, moderately sticky and plastic, moderately cohesive balls, forms long threads which will bend into rings with difficulty, moderate polish.

Sandy clay: moderately gritty, not smooth, very sticky and plastic, very cohesive balls, forms long threads which bend into rings with difficulty, high degree of polish.

Clay loam: slightly to moderately gritty, slightly smooth, moderately sticky and plastic, very cohesive balls, forms threads which will bend into rings, moderate polish.

Silt loam: not gritty to slightly gritty, very smooth and silky, slightly sticky and plastic, moderately cohesive balls, forms threads with great difficulty that have broken appearance, no polish.

Silt: not gritty to slightly gritty, extremely smooth and silky, very slightly sticky and plastic, moderately cohesive balls, forms threads with difficulty that have broken appearance, slight degree of polish.

Silty clay loam: not gritty to slightly gritty, moderately smooth and silky, moderately sticky and plastic, moderately cohesive balls, forms threads which will not bend into rings, moderate polish.

Silty clay: not gritty to slightly gritty, moderately smooth and silky, very sticky and plastic, very cohesive balls and long threads which bend into rings, high degree of polish.

Clay: not gritty to slightly gritty, not smooth, extremely sticky and plastic, extremely cohesive balls and long threads which bend into rings easily, high degree of polish.

During the determination of texture of many tropical soils they become more and more clayey as rubbing proceeds; this is known as subplasticity.

Sandy soils generally have a low water and nutrient holding capacity. Clays have a good nutrient and water holding capacity but they may become waterlogged. The very silty soils tend to erode very easily. The most desirable soils for cultivation are the loams particularly those containing about 5–10 per cent organic matter.

7. Colour

A very high proportion of the names of soils is based upon colour, since this is the most conspicuous property and sometimes the only one that is easily remembered. Further, many inferences made about soils are based upon colour.

The colour of the soil is usually determined by the nature of the fine material, the properties of which are discussed on page 93.

Generally the colour of a soil is determined by the amount and state of iron and/or organic matter. Hematite is responsible for the red colour of some soils of tropical and subtropical areas. However, the mineral responsible for most of the inorganic colouration of freely drained soils is goethite which has colours that range from reddish brown to yellow as its degree of hydration increases. The highly hydrated yellow and yellowish-brown forms are sometimes referred to as limonite.

Grey, olive and blue colours occur in the soils of wet situations and originate through the presence of iron in the reduced or ferrous state. The colour of the upper horizons usually changes from brown to dark brown to black as the organic matter content increases. Dark colours are produced also through the presence of manganese dioxide or may be caused by elemental carbon following burning.

Pale grey and white originate through the lack of alteration of light-coloured parent materials, deposition of calcium carbonate and efflorescence of salts. Pale colours also originate by the removal of colouring substances to form the distinctive leached horizons in soils such as Podzols. Another aspect of colour is that the surfaces of peds may differ from their interiors due to the presence of a coating or a bleached surface.

Some horizons have a colour pattern which may be mottled, streaked, spotted, variegated or tongued. Possibly the most common and important colour pattern is yellow and brown mottles on a grey background which is interpreted as resulting from seasonal wetting and drying of the horizon.

Although the colours of most horizons are produced by pedogenic processes there are a number of instances when they are inherited from the parent material: for example, many sediments of Devonian and Permo-Triassic age are bright red in colour.

8. Fabric of soil materials

This is the arrangement, size, shape and frequency of the individual solid soil constituents within the soil as a whole and within individual features themselves. The particles in soils range widely in size and shape so that, likewise, their mutual arrangements vary enormously. Generally the clay particles behave differently from the others mainly because of their smaller size and laminar form. The arrangement of the clay particles cannot be seen with the optical microscope but it is possible to infer certain relationships because packing and alignment of clay particles give optical anisotropism that can be seen in thin sections with the petrological microscope between crossed polars or with circularly polarised light (see next paragraph). Because of their various shapes the larger particles do not behave in a consistent manner with regard to each other or to the fine material. Sand grains usually occur singly and equidistantly but may form clusters. In a number of cases the gravel and stones may form nuclei for the development of other features such as calcite pendants (Fig. 4.3).

9. Fine material and matrix

The fine material is the mineral and/or organic material $< 2\ \mu m$ that is not easily resolved with the petrological microscope. The fine material composed of clay and/or humus is usually responsible for the soil colour which is determined by the sum of the individual constituents. The fine material ranges from thin coatings around grains to forming a complete matrix enclosing grains and pores, but matrices can be composed of coarser material. Thus the term matrix is used in a general manner for the material of any size completely surrounding and enclosing coarse material.

The clay and clay humus matrices when present in large amounts appear in plane polarised light as a continuous phase completely enclosing large grains, pores and other features. However, all the material $< 2\ \mu m$ is not always present as matrix for in some cases the fine material occurs as separate features such as clay coatings on grains or peds.

The colours of fine material matrices vary considerably: they may be yellow, brown, red or opaque.

When transparent clayey matrices are examined between crossed polars or with circularly polarised light they vary from those which are completely black or isotropic to those which show different patterns of interference colours or optical anisotropism. In isotropic matrices the particles are randomly orientated while anisotropism is caused by the alignment of the particles to form domains and lines. Domains are small anisotropic areas with pale first order colours and seem to be small groups of closely packed and similarly aligned clay particles. They are visible between crossed polars but not normally seen in plain transmitted light. Domains may be randomly distributed and orientated or they may form zones or aureoles around grains or pores (Figs 4.5 and 4.6). Anisotropic lines are long thin areas that occur within the matrix or on ped surfaces (see next para.) and seem to be the thin section manifestation of slickensides (Fig. 4.7).

10. Structure and porosity

This refers to the degree and type of aggregation and the nature and distribution of pores and pore space. In many soils the individual particles exist as discrete entities but in others the most common arrangement is for the particles to be grouped into aggregates with fairly distinctive shapes and sizes. These aggregates are known as *peds*. The range is from those very sandy soils in which each particle is separate through those with well formed aggregates to those that are compact and massive. In soils with well formed aggregates or when there is mainly sand the individual units may have only a few points of contact but generally are surrounded by a continuous pore phase. On the other hand in massive soils it is the mineral material that forms the continuum but it may and often does contain discrete pores.

The soil pore system includes discrete pores within peds and massive material and continuous pore space between peds. The size and continuity of the pore space are important for water movement and retention. The large pores ($> 10\ \mu m$) conduct water, the medium-sized pores ($10–0.2\ \mu m$) hold water which is available to plants while the fine pores ($< 0.2\ \mu m$) hold unavailable water.

Up until recently the description of structure has been based largely on hand specimens. With the development of a technique for preparing large thin sections of soils a fuller appreciation of structure is possible.

The principal types of structures are described below.

*__Alveolar:__ abundant circular, ovoid or irregular pores forming a honeycomb type of arrangement with a continuous soil phase. This type of structure is most easily identified in thin sections and may not be recognised in the field.

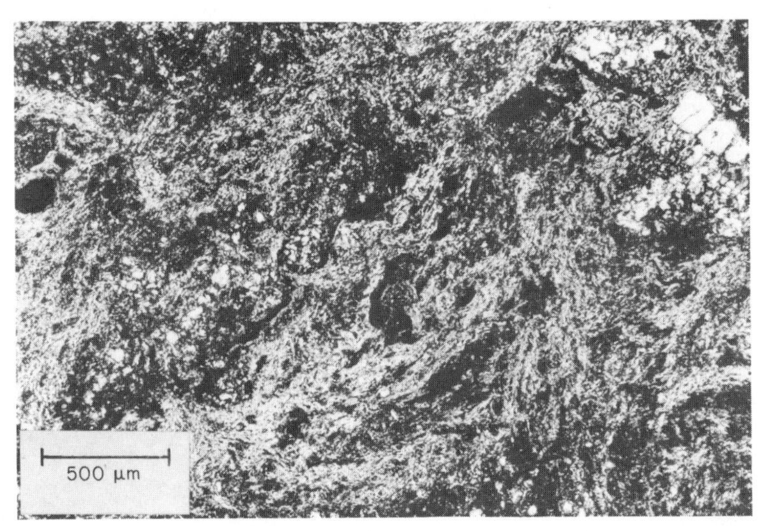

Fig. 4.5 Moderately anisotropic matrix: there are common flecked anisotropic zones as well as rare continuous zones

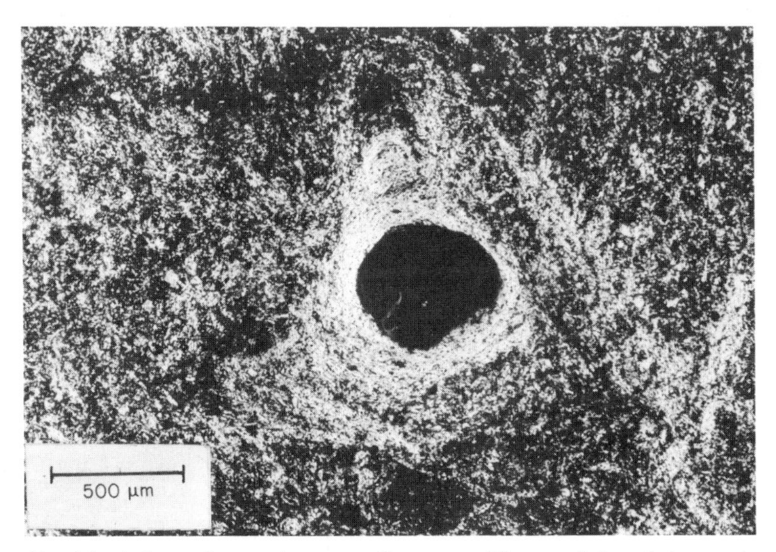

Fig. 4.6 Anisotropic aureole surrounding a pore.The remainder of the matrix is moderately anisotropic mainly with small random domains

Fig. 4.7 Numerous horizontal anisotrophic lines – circularly polarised light. These are parallel and probably result from ploughing. The remainder of the matrix is weakly anisotrophic with small random domains

Angular blocky: peds with sharp angular corners, flat, convex and/or concave faces. They are usually firm or hard and fairly tightly packed, common in many middle horizons (Figs. 4.8 and 4.15).

***Bridge:** fine material bridging single grains and forming abundant irregular pores. This is best seen in thin sections (Fig. 4.9).

***Channel:** frequent to abundant intertwining channels containing varying amounts of worm faecal material.

Columnar: large, vertically elongate peds with domed upper surfaces and three or more flat vertical faces. The peds usually grade into the underlying material and may be composed of smaller peds. This is the distinctive middle horizon of many Solonetz and Solodic Planosols (Figs 4.10 and 4.15).

Composite: various combinations of many of the other types.

Compound: large peds such as prisms or columns that are themselves composed of smaller incomplete peds.

***Crumb:** irregularly shaped peds with rough surfaces and forming a loose porous mass, common in the upper horizons beneath grassland (Figs 4.11 and 4.15).

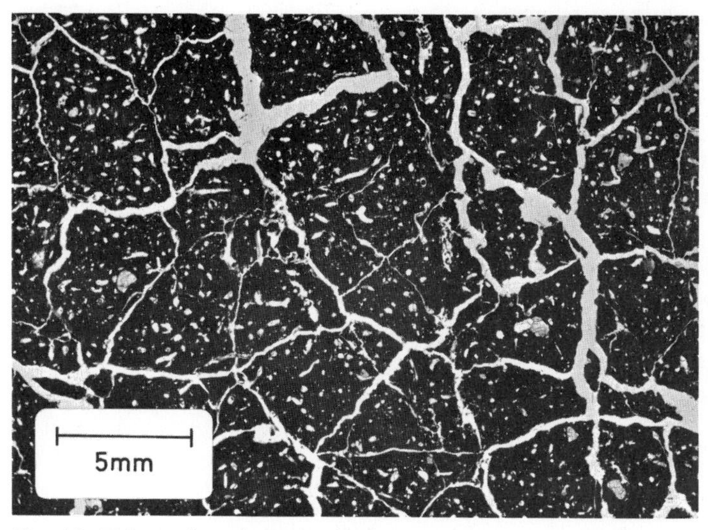

Fig. 4.8 Thin section of angular blocky structure, there is a high content of ovoid and circular pores within the individual peds

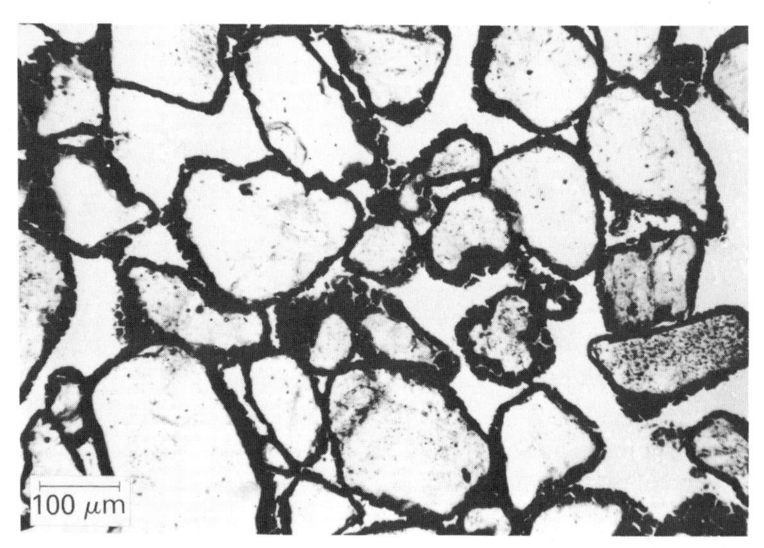

Fig. 4.9 Bridge structure formed by humus coating the sand grains and forming bridges between them

Fig. 4.10 Columnar structure – the width of the section is 1 m

Granular: subspherical peds, usually forming a fairly loose porous mass; common in many upper horizons (Figs 4.12 and 4.15).

Irregular blocky: irregularly shaped peds with flat concave and/or convex faces, mixed rounded and angular corners and re-entrant surfaces.

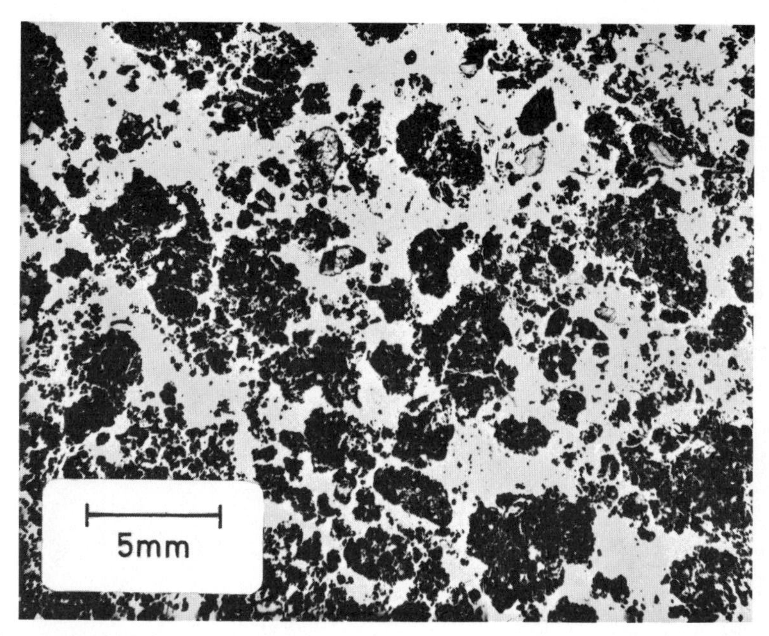

Fig. 4.11 Thin section of crumb structure composed of numerous loose porous aggregates

***Labyrinthine:** abundant intertwining faunal passages containing varying amounts of granular material. This structure is formed mainly by termites and occurs in the upper horizon of many tropical soils (Fig. 4.13).

Laminar: horizontally elongate peds with flat parallel surfaces.

Lenticular: lens-shaped peds with convex surfaces; they overlap each other and are common in many horizons that have been frozen or compacted by implements (Fig. 4.15).

Massive or coherent: continuous soil phase without peds or continuous pore space, usually found in lower horizons or underlying material (Fig. 4.15).

Prismatic: vertically elongate peds with three or more vertically flat faces. The peds often have somewhat indeterminate upper and lower boundaries. This structure is common in many medium and fine textured soils (Figs 4.14 and 4.15).

Single grain: no aggregates – individual detrital grains and small

* Best seen in thin sections.

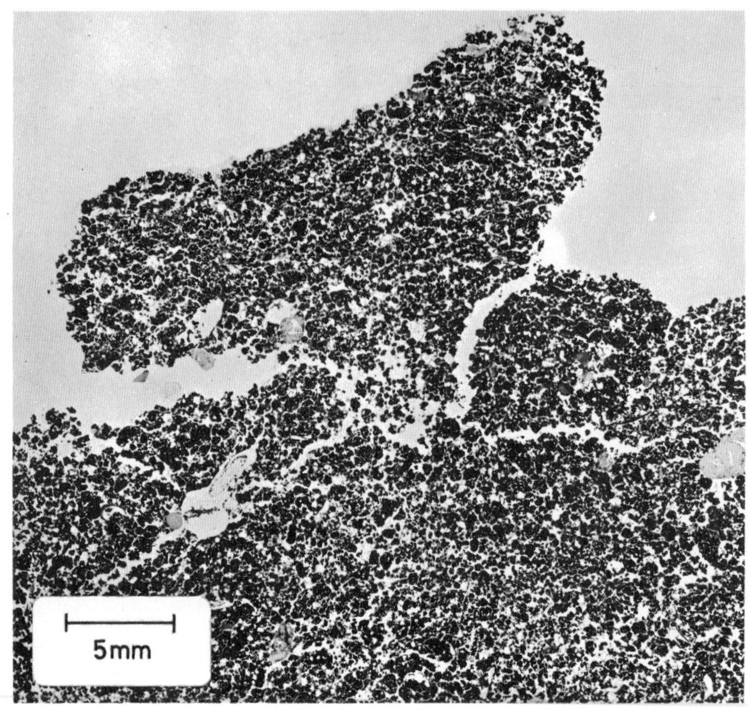

Fig. 4.12 Thin section of granular structure in the upper horizon of a Vertisol

rock fragments, common in strongly leached horizons, Arenosols and underlying unaltered sandy material.

Spongy: a tangled sponge-like mass of organic matter that usually occurs at the surface.

Subangular blocky: peds with convex and/or concave faces and rounded corners, common in many middle horizons.

Subcuboidal: peds with square or rectangular outline, angular corners and flat, strongly accordant surfaces.

Vermicular: a maze of intertwining faunal passages which are filled or partially filled with vermicular aggregates that are usually faecal in origin, common in upper horizons with vigorous earthworm activity.

Wedge: wedge shaped peds formed by the intersection of planar pores at 30°–60°. Usually occurs in middle horizons containing a large amount of montmorillonite and formed by expansion and contraction (Fig. 4.15).

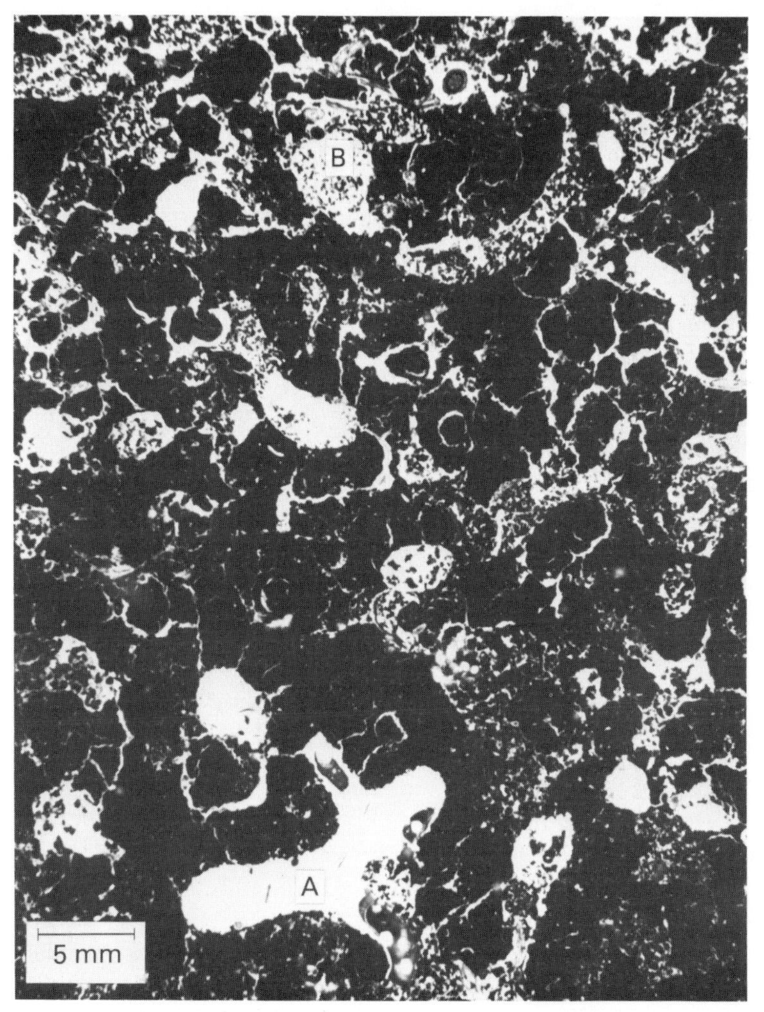

Fig. 4.13 Labyrinthine structure produced by termites. A is a termite passage and B are granules produced by termites

A diagrammatic representation of some types of structure is given in Fig. 4.15.

Structure is one of the least permanent properties of soils for it can be altered very rapidly by cultivation or any other type of disturbance. It is also a very important property, thus at present a considerable amount of research is being devoted to the

Fig. 4.14 Prismatic structure in the middle horizon. Note the sharp edges and flat faces of the prisms

formation and retention of structures such as crumb or granular. Soils with a granular or crumb structure allow free percolation of excess moisture and at the same time roots are free to grow in the pore space between the peds. Soils with massive or prismatic structure severely restrict root development because there is little pore space into which roots can grow and often they are anaerobic.

The formation of a fairly stable structure is very difficult but it can be helped by increasing the amount of organic matter, liming, drainage and reduced cultivation.

Instability in the structure often leads to puddling, crusting and

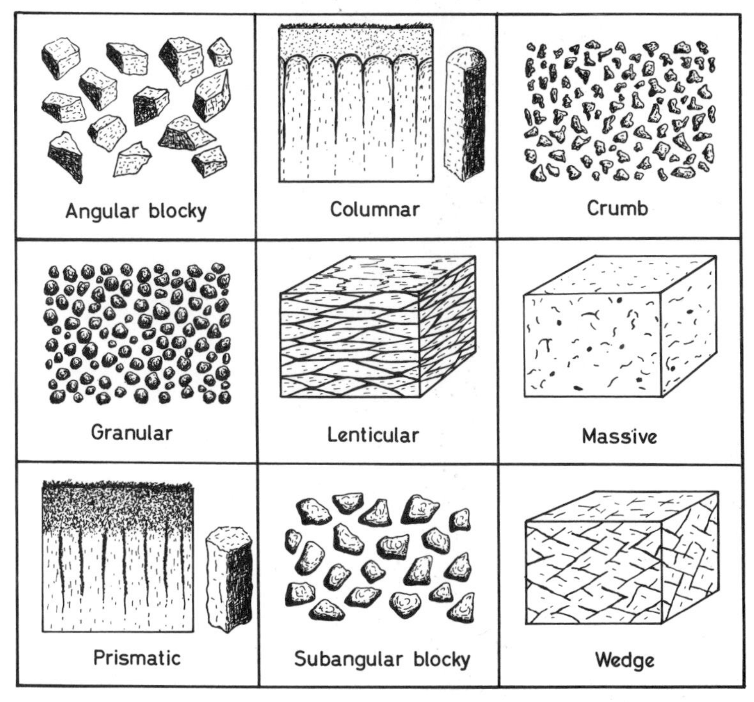

Fig. 4.15 Diagrammatic representation of some types of soil structure (not drawn to scale)

the formation of hard setting surface horizons; structure collapse also leads to the flow of soil down cracks, and these in turn lead to soil erosion.

In continuously cultivated soils in temperate regions the structure is cyclic. Starting with the well aggregated seed bed in spring, the aggregates gradually collapse during the growing season to form angular blocky peds often with earthworm vermiforms. Finally a massive structure develops during the winter.

In some tropical soils (Ferralsols) the aggregates are extremely stable and resistant to breakdown so that the soil can be ploughed immediately after a heavy shower of rain. This stability seems to result from the high content of iron compounds.

Farmers very often refer to the *tilth* of the soil. This is the physical state of the soil in relation to plant growth and is the sum total of particle size distribution, texture, structure, porosity and density. The ideal tilth required by the farmer is one of medium texture, crumb or granular structure, friable and porous.

11. Atmosphere

The soil atmosphere occupies the pores and pore spaces in soils and is generally continuous with the above ground atmosphere but differs from it in many ways. The soil atmosphere usually has a much higher content of water vapour and a much higher content of carbon dioxide which may vary from 0.25 per cent to 5 per cent as compared with 0.03 per cent in the above ground atmosphere. The atmosphere in the soil is constantly circulating and interchanging with that above ground. If it did not circulate the content of carbon dioxide and other gases would increase to proportions toxic to plant roots.

12. Passages – faunal and root

Organisms such as earthworms, termites and beetle larvae that burrow through the soil create passages. Termites produce small granules by manipulating the soil between their mandibles, earthworms and beetle larvae leave varying amounts of faecal material in their passages.

Plant roots also form passages that may become filled with material when the plant dies. Good examples are seen in semi-arid and arid areas where pipes of calcium carbonate fill the spaces left by the roots. In wet soils local oxidation around the sides of the old root passage forms tubes of material cemented by iron oxides.

13. Faecal material

This is usually recognised by its characteristic morphology. Arthropods produce small individual or clustered granular or ovoid units (Fig. 2.27). Earthworms produce vermicular (worm-like) deposits. Generally, faecal material occurs at the surface associated with the organic matter but some earthworms penetrate deep into the soil and where they are very active the soil is a tangled mass of their vermiforms.

14. Coatings

These are deposits of material on the surfaces of aggregates, pores, rock fragments and detrital grains. They include thin films of translocated clay, humus, sesquioxides, silt, calcite, opaline silica and many other substances. Thus they vary widely in their composition and are best studied in thin sections.

Clay coatings are probably the most common type and are seen in hand specimens with a hand lens to form waxy coverings on the surfaces of aggregates. In thin sections in plane polarized light they have a layered fabric and absence or low frequency of coarse material. Between crossed polars they usually show interference colours and black extinction bands (Fig. 4.16). Clay coatings are not permanent features, they become fragmented or assimilated into the matrix.

Sesquioxide coatings are common on sand grains and rock fragments in the middle horizons of Podzols. Calcite coatings are common around rock fragments in the soils of arid and semi-arid areas (Fig. 4.17). Coatings of manganese dioxide are common in soils that show some degree of wetness.

15. Moisture status

The overall annual moisture status of soils is inferred from morphological characteristics and on limited numbers of measurements made at various times of the year. On this basis the following five classes of drainage for the whole soil are usually recognised. The forms in which water occurs in soils are given on page 35.

Excessively drained

Water moves rapidly through the soil which has bright colours due to oxidising conditions.

Freely drained (Colour Plates IB & IIB)

Water moves steadily and completely through the soil with little tendency to be waterlogged. Such a soil also has bright colours due to oxidising conditions.

Fig. 4.16 Clay coating (A) in plain transmitted light showing the characteristic crescentic laminations. (B) between crossed polars showing the distinctive black extinction band

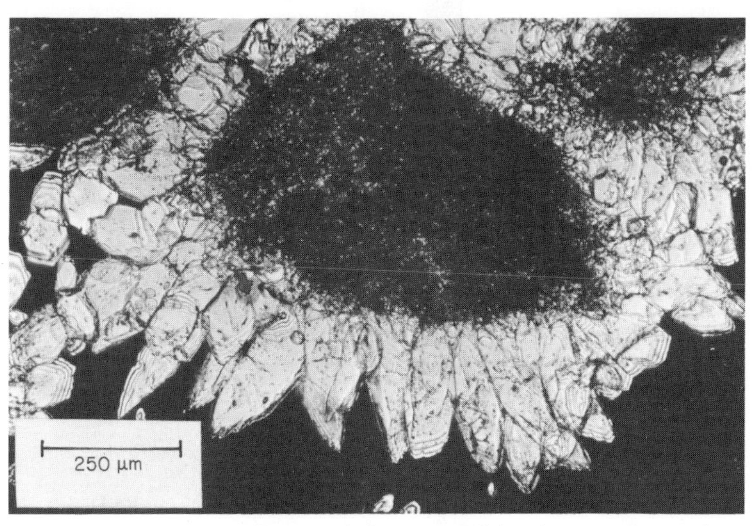

Fig. 4.17 Calcite surrounding a rock fragment and forming a pendant – circularly polarised light

Imperfectly drained

The soil is moist for part of the year with one or two horizons showing mottling owing to extended periods of wetness and reduction of iron to the ferrous state. The wetness in the soil as in the other two classes below may be caused by an impermeable horizon, high water-table or high precipitation.

Poorly drained

The soil is wet for long periods of the year with the result that many of the horizons are mottled, with at least one that is blue or grey resulting from reducing conditions.

Very poorly drained (Colour Plate IIIA)

The soil is saturated with water for the greater part of the year so that most of the horizons are blue or grey caused by reducing conditions. Peat may also be present as a consequence of the high degree of wetness.

The water in drains and small streams in areas of wet soils often shows rusty colours and glistens. The former is due to ferric hydroxide and the latter to aluminium hydroxide.

Periodic submergence causes a rise in the pH of acid soils and a lowering of the pH of alkaline soils. This is especially important in pyrite containing soils that become very acid when they dry out, due to the oxidation of the pyrite to sulphuric acid. Submergence increases the content of soluble material including silica and causes the reduction of some substances, particularly compounds of iron, manganese and nitrogen. It also decreases the rate of organic matter decomposition by inhibiting the micro-organisms.

16. Density

The true density of a soil is a measure of the density of the constituent components and varies from about 2.65 for the mineral particles to about 0.2 for the organic matter. On the other hand the density of the whole undisturbed soil or *bulk density* is

$$\frac{\text{Wt of a dry block of soil (g)}}{\text{Vol of block when sampled (cm}^3\text{)}}$$

This takes into account the density of the soil materials themselves and their arrangement or structure. Therefore a loose porous soil will have a smaller bulk density than a compact soil even though the density of the individual particles in the two soils may be the same.

Whereas the bulk density of most cultivated soils is about 1.3 the extremes can vary from 0.55 for soils developed on volcanic ash to 2.0 for some strongly compacted lower horizons. This property is attaining some importance in fertility studies because continuous cultivation by heavy implements increases the bulk density by inducing compaction which reduces percolation and root penetration.

17. pH

The pH of a solution is a measure of its acidity or alkalinity and is defined as the negative logarithm of the hydrogen ion concentration. Soils do not behave like simple solutions, therefore it is not possible to give an accurate definition of soil pH but for many purposes it can be considered as similar to the above. The range for soils is normally from pH 3 to 9. Very low values develop following the drainage of coastal marshes and swamps that contain pyrite which is oxidised, resulting in the production of sulphuric acid.

At the other extreme very high values result from the presence of sodium carbonate. Within the normal range the two principal controlling factors are organic matter and the type and amount of cations. Large amounts of organic matter induce acidity except when counterbalanced by high concentrations of basic cations. Acidity is also induced by large amounts of aluminium in solution. Generally pH values about neutrality are associated with large amounts of exchangeable calcium and some magnesium, sometimes supplemented by free carbonates. As discussed later, pH is a very important factor in plant growth. Where laboratory facilities are limited pH can sometimes be used to determine the lime requirement and can be estimated easily in the field and therefore can be used as a mapping criterion.

A simple classification of pH is as follows:

Extremely acid	<4.5
Very strongly acid	4.5–5.0
Strongly acid	5.0–5.5
Moderately acid	5.5–6.0
Slightly acid	6.0–6.5
Neutral	6.5–7.3
Slightly alkaline	7.3–7.8
Moderately alkaline	7.8–8.5
Strongly alkaline	8.5–9.0
Very strongly alkaline	>9.0

18. Organic matter

Up to the present the exact nature of the organic matter in soils has not been determined. Therefore in routine analyses of soils only the total amounts of carbon and nitrogen are determined.

The carbon percentage is usually multiplied by the conversion factor of 1.72 to give an indication of the total amount of organic matter present. The ratio of carbon to nitrogen (C/N ratio) is calculated and used as a measure of humification and ranges from >100 to 8. The former high figure is for fresh litter and peat while the latter occurs in many upper horizons containing a mixture of mineral material and well humified organic matter.

A rough measure of the amount of organic matter in soils can be obtained by igniting dry soil and determining the loss in weight. This technique is applicable to most soils except those containing carbonates which decompose when ignited.

The total amount of organic matter in soils varies from <1 per cent to >90 per cent. The latter figure is for the relatively unaltered material that occurs in the litter at the surface and in peat. Normally the upper horizons contain <15 per cent organic matter and over very large areas they contain <2 per cent, even when the supply of litter to the surface is high as in certain humid tropical areas.

Plant material is composed mainly of carbohydrates, proteins and fats with smaller amounts of lignins, waxes and resins. The carbohydrates are composed of sugars, starch, hemi-cellulose and cellulose, that are decomposed by the soil fauna and flora to provide their energy. Proteins are decomposed to amino acids mainly to provide body tissue of the organisms and also for the various nitrogen transformations that take place. These processes cause a gradual narrowing of the C/N ratio. The fats, waxes, resins and lignins are decomposed slowly, thus remaining in the soil for some time.

Organic matter is rapidly lost by cultivation; there can be as much as 50 per cent loss of organic materials in humid tropical conditions within 10 years from 4 to 2 per cent. Probably the most dramatic losses occur in some amorphous peats where the surface may be lowered by about 1–2 m over a prolonged period of cultivation. In mineral soils there is unlikely to be a complete loss because of the supply of plant residues and dead crop roots. In addition there is 'protected' organic matter, strongly linked to the clays and probably occurring in pores too small for the entry of bacteria. This organic matter is very old, probably thousands of years.

The decomposed organic matter that is well mixed with the mineral material in the upper horizons is the *humus*.

The true character of humus is still uncertain but it is now generally believed to be a product of plant decomposition and microbiological synthesis and composed of large molecules of heterogeneous polymers formed by the interaction of polyphenols, amino acids, polysaccharides and other substances. The former two are mainly products of plant decomposition while the polysaccharides are products of microbiological synthesis. Humus has unique properties which very largely determine the character of the upper horizons. Firstly, humus is capable of absorbing large quantities of water thus increasing the water holding capacity of the soil and is therefore of importance in crop production. Five per cent humus will increase the water holding capacity of a sandy loam by more than 50 per cent and a clay loam by 30 per cent. Like clays humus has a Cation Exchange Capacity (CEC), but it is considerably higher, being about 3000 meq kg^{-1} and therefore increases considerably the cation holding capacity of the soil. Humus can be dispersed or flocculated depending upon the nature of the cations present and, as already stated on page 84, it influences handling consistence. Thus humus behaves somewhat like clay but it is easily destroyed by microorganisms, hence the difficulties encountered in keeping it at a high enough level in many soils. In addition it has already been stated that organic matter affects soil colour and it supplies essential elements when it is decomposed.

Some organic compounds such as the polyphenols form soluble complexes with some metal elements such as iron and move downwards in the soil solution. Under natural conditions the humus content of a virgin soil is usually higher than in adjacent cultivated areas. This is caused by a higher rate of addition of organic matter by the natural vegetation accompanied by a lower rate of biological activity and lower temperatures. The reduction in the humus content of the soil is probably the main cause of the deterioration of structure, reduced water and nutrient availability and cultivation problems.

The classical approach to the study of soil organic matter is to treat the soil with dilute sodium hydroxide (0.1 to 0.5 M) which extracts a part of the organic matter, then to acidify the filtrate which causes part of the extracted organic matter to be precipitated. The part that stays in solution is generally known as fulvic acid and that which is precipitated as humic acid, the unextracted residue being the humin. The humic acid precipitate is further fractionated by treatment with ethanol which dissolves a part of the precipitate known as hymanomelanic acid and the remainder the insoluble humic acid.

Although the fractions are given names they are not homogeneous for each contains substances with a wide range of molecular weights. The molecular weight of fulvic acid is generally below 10 000 and humic acid above 5000 and going up to several million.

The importance and value of organic matter has led to the extreme idea of 'organic farming', i.e. the use of organic matter to maintain soil fertility with the total exclusion of chemical fertilizers. Although the idea is intrinsically good it will not generally produce high crop yields.

A summary of the benefits of organic matter is as follows:

(a) Improves structure and structure stability
(b) Increases water holding capacity
(c) Increases CEC
(d) Improves conditions for microbial growth
(e) Nutrient reserve
(f) Decreases toxicity of Al
(g) Improves tilth
(h) Adsorption/deactivation of organic pesticides

19. Cation exchange properties

The two most important cation exchange properties are the cation exchange capacity (CEC) and the percentage base saturation.

The CEC of the whole soil is a measure of the exchange capacity or negative charges of the clay and humus expressed as milligram equivalents per 1000 grams of soil i.e. meq kg^{-1}. The range is from about <50 meq kg^{-1} for some lower horizons up to >1000 meq kg^{-1} for upper horizons containing high percentages of organic matter, vermiculite or montmorillonite.

The percentage base saturation is a measure of the extent to which the exchange complex is saturated with basic cations. The general trend is for the amount of exchangeable bases to increase with decreasing rainfall and for calcium to be dominant but sodium may be dominant in certain arid regions. Conversely low figures indicate intense leaching but changes can be induced very easily by cultivation. Continuous cultivation leads to a rapid reduction in cations such as calcium and potassium if they are not added in fertilizers.

When the total content of exchangeable cations is expressed as meq kg^{-1} the amount may seem small but when recalculated on the basis of the amounts of cations available to a growing crop there are usually several thousand kg of cations per hectare.

The negative charges in soils are of two types: permanent charges and pH dependent charges. The permanent charges are of three types:

1. Internal charges due to isomorphous replacement of Al and Mg in the octahedral layer and are highest in the 2:1 types of clays.
2. Broken bonds resulting from fracture of the minerals
3. Ionisation or dissociation of H^+ from OH at the edges of the clay lattice leaving 0^- with a negative charge.

The pH dependent charges are directly related to soil pH. At low pH values, the charge and thus the CEC is low but increases as the pH value rises.

20. Soluble salts

When a soil is shaken with water and filtered, the filtrate will contain some dissolved salts but they only occur in significant proportions in the soils of arid and semi-arid areas (see page 194). There they accumulate because the annual precipitation is insufficient to leach the soils or because the water-table is at a shallow depth and moisture is drawn to the surface by capillarity bringing with it dissolved salts which are left behind as the moisture evaporates (Fig. 4.18). Flooding by sea water also causes salinity in soil but this is of minor importance except in countries such as Holland which depend upon large areas reclaimed from the sea. The predominant anions are bicarbonate, carbonate, sulphate and chloride while the cations include sodium, calcium, magnesium and small amounts of potassium.

These ions occur in widely varying proportions and depending upon the particular ratio they impart a number of properties to the soil, some of which are detrimental to plant growth. Soils can be grouped into two scales depending upon their salinity and alkalinity.

Fig. 4.18 Salt efflorescence on the surface of the soil and a partly exposed Solonchak profile

DEGREE OF SALINITY	% by weight
Free of salts	<0.15%
Slightly saline	0.15–0.35%
Moderately saline	0.35–0.65%
Strongly saline	>0.65%

DEGREE OF ALKALINITY	% of total exchangeable ions
Slightly alkaline	<20% exchangeable sodium
Moderately alkaline	20–50% exchangeable sodium
Strongly alkaline	>50% exchangeable sodium

A common feature of many soils containing soluble salts is that microelements accumulate to toxic proportions. In the case of boron concentrations over 1.0 μg/g are toxic. Generally plants cannot tolerate a salt concentration of over 0.5 per cent and as the concentration rises much above this limit the percentage of halophytes such as *Atriplex* spp and *Beta* spp increases rapidly.

There are immense areas of potentially very fertile land that have a high concentration of salts but in many cases the salts can be removed, rendering the land suitable for agriculture as discussed on page 172.

21. Carbonates

Carbonates of calcium and magnesium, particularly the former, are widely distributed in soils, occurring separately or they may be associated with other salts. The most important properties of carbonates are: (1) They are relatively easily soluble in water containing dissolved carbon dioxide, and therefore can be quickly lost or redistributed within the soil. (2) When present in an amount as small as 1 per cent of the soil they can dominate the course of soil development because this amount is sufficient to raise the pH value over neutrality and sustain a high level of biological activity. (3) Carbonates, particularly calcium carbonate, are the first substances to start accumulating as the climate becomes arid. (4) Both calcium and magnesium are essential plant nutrients. (5) Carbonates are regularly added to many arable soils to raise their pH values for optimum plant growth.

22. Elemental composition

The principal elements occurring in soils and their proportions in the Earth's crust have already been given (page 23) and only in very detailed studies is it necessary to determine the total amount of each element present. Where soil formation has proceeded for a long period and where there has been a considerable amount of hydrolysis and solution, it is usual to perform a quantitative analysis to estimate the ten or so dominant elements in each horizon as well as in the underlying unaltered material. This is to determine the predominant chemical changes that have taken place during soil formation. The elements which are determined include silicon, aluminium, iron, sodium, potassium, calcium, magnesium, manganese, titanium, zirconium, nitrogen and phosphorus. However, the determination of the total amount of certain elements is performed very frequently, particularly nitrogen and phosphorus because of their importance as essential plant nutrients.

23. Amorphous oxides – free silica, alumina and iron oxides

The type, amount and distribution of amorphous materials can be used as criteria for measuring the degree and type of soil formation. Also they are regarded as being formed during the current phase of pedogenesis and, therefore, are valuable criteria for differentiating between relic and contemporary phenomena. When perfoming the determinations it is customary to treat the soil with an extracting reagent such as sodium dithionite, citric acid or sodium pyrophosphate and to determine the amount of each constituent in the extract. Published data pertaining to these techniques can be challenged but certain trends have been discovered which seem to be reasonably valid. Some of these trends are given in Chapter 7.

Although chemical techniques have been used almost exclusively for the study of these oxides, other methods, including the study of soils in thin sections, by the electron microscope and electron probe, are yielding many new and interesting data. For example, it has been shown in some instances that silica often forms small aggregates while alumina and iron oxides form coatings on clay minerals and other surfaces.

24. Concretions

Under certain conditions some constituents form local concentrations which may become very hard. For example, iron may accumulate to form concretions in certain soils of the tropics and some periodically waterlogged soils. Similarly, calcium carbonate and gypsum form concretions or massive horizons in arid and semi-arid areas.

25. Engineering soil properties

Some engineering properties are determined by elaborate apparatus but many can be made with inexpensive equipment.

The properties of soils usually required by engineers include: particle size distribution, bulk density, porosity, permeability, bearing capacity, expansion and contraction, cohesion, shearing, compressibility and mineralogy. Also included are the Atterburg limits, viz. the liquid limit, the plastic limit and the plasticity index. The liquid limit and the plastic limit are determined by special laboratory techniques and defined as follows.

The liquid limit of a soil is the amount of water present at the point at which the soil changes from the plastic to the liquid state and expressed as a percentage of the oven dry soil.

The plastic limit of a soil is the smallest amount of water at which the soil remains plastic and expressed as a percentage of the oven dry soil. The plasticity index is the difference between the liquid limit and the plastic limit. Some materials such as those with coarse textures are not plastic therefore they do not have the above properties.

Expansion and contraction or the shrink-swell potential determines the volume change with differences in moisture and is measured by the coefficient of linear extensibility or COLE. This is the change in length of the soil upon drying:

$$\frac{Lm - Ld}{Ld}$$

where Lm = length of the moist sample
Ld = length of dry sample.

This is a very useful determination for assessing the value of soils for foundations, roads, canals, etc.

At present there is a growing tendency for soil surveyors to include engineering properties in their field observations and also to collect samples for the determination of engineering properties. Thus many modern soil maps give useful engineering data. However engineers have their own classification.

Another important engineering property of soils is their corrosivity which generally increases with acidity. In addition wet, fine textured, saline soils are also very corrosive.

Corrosivity for concrete also increases with decreasing pH and

Table 4.2 Typical names, group symbols and characteristics of soil materials of the Unified soil classification system (USDA 1971)

Group symbol	*Typical names*	*Compressibility*	*Compaction characteristics*	*Permeability of compacted soil*
GW	Well-graded gravel, gravel and sand mixtures, little or no fines	Low	Good	High
GP	Poorly graded gravel, gravel and sand mixtures, little or no fines	Low	Good	High
GM	Silty gravel, gravel and sand and silt mixtures	Low	Fair to good	Medium to low
GC	Clayey gravel, gravel and sand and clay mixtures	Low to medium	Good to fair	Low
SW	Well-graded sands, gravelly sands, little or no fines	Low	Good	High
SP	Poorly graded sands, gravelly sands, little or no fines	Low	Good	High
SM	Silty sands, sand and silt mixtures	Low to medium	Fair to good	Medium to low
SC	Clayey sands, sand and clay mixtures	Low to medium	Good to fair	Low
ML	Inorganic silts and very fine sands, rock flour, silty or clayey fine sands, or clayey silts with slight plasticity	Medium	Fair to good	Medium to low
CL	Inorganic clays of low to medium plasticity, gravelly clays, sandy clays, silty clays, lean clays	Medium	Fair to good	Low
OL	Organic silts and organic silty clays of low plasticity	High	Poor	Low to medium
MH	Inorganic silts, micaceous or diatomaceous fine sandy or silty soils, elastic silts	High	Fair to poor	Low
CH	Inorganic clays of high plasticity, fat clays	High	Fair to poor	Low to medium
OH	Organic clays of medium to high plasticity, organic silts	High	Poor	Low
Pt	Peat, muck, and other highly organic soils	Not suitable	Not suitable	Not suitable

increasing amounts of water soluble sulphates. For steel low corrosivity is about neutrality and low salinity.

Engineers use the unified soil classification system to classify the suitability of soils for roads, embankments, foundations and airfields. There are fifteen classes based largely on particle size distribution. They are divided into gravels (G), and sands (S), which are further subdivided on the basis of the associated material into well graded material (W) and poorly graded material (P). The fine textured material associated with the gravel and sand and fine textured soils themselves are divided into silt (M) and clay (C) depending upon the liquid limit and plasticity index and subdivided into those with low (L) or high (H) liquid limits. The silt and clay groups are also subdivided on their content of organic matter into OL and OH. Mainly organic materials are designated Pt. Table 4.2 shows the fifteen categories and some of their properties.

5 Horizon nomenclature

There is as yet no internationally agreed system of soil classification and soil horizon nomenclature. However the FAO system comes close to being the general method for communicating information and will be used in this book but where necessary equivalents as used by the USDA (1975) and FitzPatrick (1980) will be given.

Soil horizon designation

Capital letters are used to indicate master horizons and a combination of letters is used for transitional situations. Lower case letters are used as suffixes to qualify the master horizons and Arabic figures are used as suffixes to indicate vertical subdivision of a horizon. In the case of A and B horizons, lower case suffix always precedes the suffix figure. Arabic figures are also used as prefixes to indicate lithological discontinuities.

Master horizons

H: Formed by the accumulation of organic material deposited on the surface and is saturated with water for prolonged periods. Generally it contains 20 per cent or more organic matter.

O: Formed by the accumulations of organic material deposited on the surface that is not saturated with water for more than a few days of a year and contains 35 per cent or more of organic matter.

A: A mineral horizon at or adjacent to the surface having a morphology acquired by soil formation and usually shows an accumulation of humified organic matter intimately associated with the mineral fraction.

E: A pale coloured mineral horizon showing a concentration of sand and silt fractions high in resistant minerals resulting from the loss of silicate clay, iron or aluminium or some combinations of them. E horizons are eluvial horizons which generally underlie an H, O or A horizon and overlie B horizons.

B: A mineral horizon in which parent material is absent or faintly evident and characterized by one or more of the following features:
  (a) an illuvial concentration of clay minerals, iron, aluminium, or humus, alone or in combinations;
  (b) a residual concentration of sesquioxides relative to the source materials;
  (c) an alteration of material from its original condition to the extent that silicate clays are formed, oxides are liberated, or both, or granular, blocky, or prismatic structure is formed.

  B horizons differ greatly and generally need to be qualified by a suffix to have sufficient connotation in a profile description. A "humus B" horizon is designated as Bh, an "iron B" as Bs, a "textural B" as Bt, a "colour B" as Bw.

C: A mineral layer of unconsolidated material similar to the material from which the solum is presumed to have formed and which does not show properties diagnostic of any other master horizons. Strictly this is not a horizon.

  Accumulations of carbonates, gypsum or other more soluble salts may be included in C horizons if the material is otherwise little affected by the processes which contributed to the formation of these interbedded layers. When a C horizon consists mainly of sedimentary rocks such as shales, marls, siltstones or sandstones, which are sufficiently dense and coherent to permit little penetration of plant roots but can still be dug with a spade, the C horizon is qualified by the suffix m for compaction.

R: A layer of continuous indurated rock. The rock of R layers is sufficiently coherent when moist to make hand digging with a spade impracticable. Like the C this is not a horizon.

Transition horizons

Soil horizons in which the properties of two master horizons merge are indicated by the combination of two capital letters (for example AE). The first letter marks the master horizon to which the transitional horizon is most similar. Horizons that contain mixtures of two horizons that are identifiable as different master horizons are designated by two capital letters separated by a solidus (for example E/B).

Letter suffixes

A small letter may be added to the capital letter to qualify the master horizon designation. Suffix letters can be combined but normally no more than two suffixes should be used in combination (for example, Ahz).

The suffix letters used to qualify the master horizons are as follows:

b. Buried or bisequal soil horizon (for example, Btb).
c. Accumulation in concretionary form; this suffix is commonly used in combination with another which indicates the nature of the concretionary material (for example, Bck, Ccs).
g. Mottling reflecting variations in oxidation and reduction (for example, Bg, Btg, Cg).
h. Accumulation of organic matter in mineral horizons (for example, Ah, Bh); and applied only where there has been no disturbance or mixing by ploughing.
k. Accumulation of calcium carbonate.
m. Strongly cemented, consolidated, indurated; and used in combination with another indicating the cementing material (for example, Cmk marking a petrocalcic horizon).
n. Accumulation of sodium (for example, Btn).
p. Disturbed by ploughing or other tillage practices (for example, Ap).
q. Accumulation of silica (Cmq, marking a silcrete layer in a C horizon).
r. Strong reduction as a result of groundwater influence (for example, Cr).
s. Accumulation of sesquioxides (for example, Bs).
t. Illuvial accumulation of clay (for example, Bt).
u. Unspecified; used in connexion with A and B horizons which are not qualified by another suffix but have to be subdivided vertically by figure suffixes (for example, Au1, Au2, Bu1, Bu2).

w. Alteration *in situ* as reflected by clay content, colour, structure (for example, Bw).
x. Occurrence of a fragipan (for example Btx).
y. Accumulation of gypsum (for example, Cy).
z. Accumulation of salts more soluble than gypsum (for example, Az or Ahz).

Figure suffixes

Horizons designated by a single combination of letter symbols can be vertically subdivided by numbering each subdivision consecutively, starting at the top of the horizon (for example Bt1 – Bt2 – Bt3 – Bt4).

Figure prefixes

Arabic numerals are prefixed to the horizon designations to distinguish lithological discontinuities; for instance, when the C horizon is different from the material in which the soil is presumed to have formed the following soil sequence could be given: A, B, 2C. Strongly contrasting layers within the C material could be shown as an A, B, C, 2C, 3C . . . sequence.

Diagnostic horizons

Soil horizons that have a set of quantitively defined properties which are used for identifying soil units are called "diagnostic horizons". Since the characteristics of soil horizons are produced by soil forming processes, the use of diagnostic horizons for separating soil units ensures that the classification system is based on general principles of soil genesis.

Agric horizon: a dark horizon that occurs immediately beneath the plough layer. It is characterised by having >15 per cent thick coating of humus and clay.

Albic E horizon (Colour Plate IVA): is one from which clay and free iron oxides have been removed, or in which the oxides have been segregated to the extent that the colour of the horizon is determined by the colour of the primary sand and silt particles. It has a colour value moist of 4 or more, or a value dry of 5 or more, or both.

Anthropic epipedon: an upper horizon similar to the mollic A horizon but contains >250 ppm of citric acid soluble P_2O_5 and is largely the result of human activity.

Argillic B horizon (Colour Plate IIIB): is one that contains illuvial clay and has the following properties:

1. The clay increase is generally more than 3 per cent for sands, 8 per cent for clays and 1.2 times the content of the eluvial horizon for loams; and occurs within a vertical distance of 30 cm or less.
2. It has clay coatings on ped surfaces and in the pores, or has orientated clays in 1 per cent or more of the cross-section.
3. If the B horizon is clayey with 2:1 layer clays, clay coatings may be absent, provided there is evidence of pressure caused by swelling.

Calcic horizon (Colour Plate IIA): consists of secondary carbonate enrichment over a thickness of 15 cm or more, has a calcium carbonate equivalent content of 15 per cent or more and at least 5 per cent greater than that of the C horizon.

Cambic B horizon (Colour Plate IB): is an altered horizon without the diagnostic properties of other horizons and has one or more of the following properties:

1. Texture of very fine sand, loamy very fine sand or finer.
2. Soil structure or absence of rock structure in at least half the volume of the horizon.
3. Significant amounts of weatherable minerals.
4. Higher clay content and stronger chroma or redder hue than the underlying horizon.
5. Evidence of removal of carbonates or of reduction processes.

Duripan: a middle or lower horizon that is at least 50 per cent cemented by silica and does not slake in water or HCl.

Fragipan (Colour Plate IVA): hard, loamy, lower horizon with high bulk density and slakes in water.

Gypsic horizon: is a horizon of secondary calcium sulphate enrichment that is more than 15 cm thick, has at least 5 per cent more gypsum than the underlying C horizon.

Histic H horizon: is an H horizon which is generally more than 20 cm but less than 40 cm thick. A histic H horizon is eutric when it has a pH of 5.5 or more and dystric when the pH is <5.5.

Mollic A horizon (Colour Plate IIA): has the following properties:

1. Not massive or hard when dry.
2. Colours with a chroma and value of less than 3.5 when moist.
3. The base saturation is 50 per cent or more.
4. At least 1 per cent organic matter. The upper limit of organic carbon content of the mollic A horizon is the lower limit of the histic H horizon.
5. Thickness of 10 cm or more if resting on rock or a hard horizon, at least 18 cm and one third of the thickness of the solum up to 75 cm thick and more when the solum is >75 cm.
6. The content of P_2O_5 soluble in 1 per cent citric acid is less than 250 ppm; higher contents indicate anthropic influences.

Natric B horizon: has the properties of the argillic B horizon and in addition it has:

1. A columnar or prismatic structure or a blocky structure with tongues of an eluvial horizon.
2. Exchangeable sodium of more than 15 per cent within the upper 40 cm of the horizon; or more exchangeable magnesium plus sodium than calcium plus exchange acidity.

Ochric A horizon (Colour Plate IIB): is one that is too light in colour, has too high a chroma, too little organic matter, or is too thin to be mollic or umbric, or is both hard and massive when dry.

In separating Yermosols from Xerosols a distinction is made between *very weak* and *weak* ochric A horizons:

1. A very weak ochric A horizon has a weighted average percentage of organic matter less than 1 per cent in the surface 40 cm.
2. A weak ochric A horizon has a content of organic matter intermediate between the very weak ochric A horizon and the mollic A horizon.

Oxic B horizon (Colour Plate IIB): is not argillic or natric and:

1. Is at least 30 cm thick.
2. Has a cation-exchange capacity of the fine earth fraction of $\leqslant$160 meq kg^{-1} clay.
3. Has only traces of primary aluminosilicates such as feldspars and ferromagnesian minerals.
4. Has more than 15 per cent clay.

5. Has mostly gradual or diffuse boundaries between its subhorizons.

Petrocalcic horizon: a hard and indurated calcic horizon >50 per cent of which will dissolve in acid but does not slake in water. Some silica may be present.

Petrogypsic horizon: a hard horizon that usually contains >60 per cent gypsum.

Placic horizon: a thin black or dark reddish brown hard pan cemented by iron, iron and manganese or by an iron-organic matter complex. It ranges from 2 mm to 10 mm thick and has a pronounced wavy or convolute form.

Plaggen epipedon: an upper horizon that is >50 cm thick and formed by the continuous addition of soil material to the surface by humans.

Salic horizon: a horizon containing at least 2 per cent salts.

Sombric horizon: a dark-coloured middle horizon containing illuvial humus but does not have associated aluminium as in spodic B horizons or sodium as in natric B horizons. It has a low cation-exchange capacity and base saturation and seems to occur in the cool moist upland areas in the tropics and subtropics.

Spodic B horizon (Colour Plate IVA): meets one or more of the following requirements:

1. A subhorizon more than 2.5 cm thick that is continuously cemented by a combination of organic matter with iron or aluminium or with both.
2. A sandy or coarse-loamy texture with distinct dark granules.
3. Brown, dark brown or black with significant amounts of extractable carbon and/or iron and/or aluminium.

Sulphuric horizon: forms as a result of drainage and oxidation of mineral or organic materials rich in sulphides. It is characterised by a pH less than 3.5 and jarosite mottles with a hue of 2.5 Y or more and a chroma of 6 or more.

Umbric A horizon: the requirements are comparable to those of the mollic A horizon in colour, organic matter and phosphorus content, consistency, structure and thickness; but it has a base saturation of less than 50 per cent.

Letter suffixes can be used to describe diagnostic horizons and features in a profile. For example:

argillic B horizon Bt petrocalcic horizon mk

calcic horizon	k	petroferric horizon	ms
cambic B horizon	Bw	petrogypsic horizon	my
fragipan	x	plinthite	sq
gypsic horizon	y	spodic B horizon	Bhs, Bh or Bs
mottled layers	g	gleyic horizon	g
natric B horizon	Btn	strongly reduced	r
oxic B horizon	Bws		

Other diagnostic properties

A number of characteristics used to separate soil units are not horizons. They are diagnostic features of horizons or soils and are given below.

Abrupt textural change: is a considerable increase in clay content within a very short distance at the contact between an A or E horizon and the underlying horizon.

Albic materials: are exclusive of E horizons, with a colour value moist of 4 or more, dry of 5 or more, or both.

Aridic moisture regime: used to characterise Yermosols and Xerosols and to separate them from soils with a similar morphology that occur outside arid areas.

In most years these soils have no available water in any part of the soil for more than half of the time (cumulative) and the temperature at 50 cm is above 5° C.

Exchange complex dominated by amorphous material:

1. Cation-exchange capacity is more than 1500 meq kg^{-1} clay, and commonly more than 5000 meq.
2. Bulk density of the fine earth fraction is less than 0.85 g cm^{-3} at 33.3 kPa tension.

Ferralic properties: having a cation-exchange capacity of <240 meq kg^{-1} clay (Cambisols and Arenosols).

Ferric properties: many coarse red mottles, discrete iron nodules and a CEC of <240 meq kg^{-1} clay (Acrisols and Luvisols).

Gilgai microrelief: typical of clayey soils that have a high coefficient of expansion with distinct seasonal changes in moisture content. This microrelief consists of either a succession of enclosed microbasins and microknolls in nearly level areas, or of microvalleys and microridges that run up and down the slope. The height of the microridges commonly ranges from a few centimetres to 1 m.

High organic matter content in the B horizon: generally >1.35 per cent organic matter down to 100 cm or >1.5 per cent in the upper part of the B horizon.

High salinity: an electrical conductivity of the saturation extract of more than 0.15 S m^{-1} at 25° C.

Hydromorphic properties: refers to one or more of the following:

1. Saturated by groundwater, continuously or at some period of the year with evidence of reduction.
2. Occurrence of a histic H horizon.
3. Dominant hues are neutral N or bluer than 10 Y.
4. Mottling due to the segregation of iron.
5. Iron-manganese concretions >2 mm and dominant hues of 2.5 Y and 5 Y in the matrix.
6. Mottling in an albic E horizon or in the top of the spodic B horizon if both are present.
7. Presence of a thin iron pan above a fragipan or on a spodic B or with an albic E horizon underlain by a spodic B horizon.
8. Continuous plinthite within 30 cm of the surface.

Interfingering: thin (<5 mm wide) penetrations of an albic E horizon into an underlying argillic or natric B horizon along ped faces.

Permafrost: is a layer in which the temperature is perennially at or below 0°C.

Plinthite: is an iron-rich, humus-poor mixture of clay with quartz and other diluents, which commonly occurs as red mottles, usually in platy, polygonal or reticulate patterns, and which changes irreversibly to an ironstone hardpan or to irregular aggregates on exposure to repeated wetting and drying. In a moist soil, plinthite is usually firm but it can be cut with a spade. When irreversibly hardened the material is no longer considered plinthite but is called *ironstone*.

Slickensides: polished and grooved surfaces that are produced by one mass sliding past another.

Smeary consistence: characteristic of thixotropic soil material, that is, material that changes under pressure or by rubbing, from a plastic solid into a liquified stage and back to the solid condition. In the liquified stage the material skids or "smears" between the fingers (Andosols).

Soft powdery lime: refers to translocated lime soft enough to be cut readily with a finger nail.

Sulphidic materials: waterlogged mineral or organic soil materials containing 0.75 per cent or more sulphur, mostly in the form of sulphides, and having less than three times as much carbonate as sulphur. Sulphidic material differs from the sulphuric horizon in that it does not show jarosite mottles.

Takyric features: soils have a fine texture, crack into polygonal elements when dry and form a platy or massive surface crust.

Tonguing: penetration (>5 mm wide) of an albic E horizon into an argillic B horizon along ped surfaces.

Vertic properties: at some period in most years the soil shows cracks that are 1 cm or more wide within 50 cm of the upper boundary of the B horizon and usually extend to the surface.

Weatherable minerals: include nearly all 2:1 layer clays and silt and sand-size feldspars, ferromagnesian minerals, glasses and micas.

6 Soil fertility and land use

What is a fertile soil? This is a difficult question to answer since the requirements of plants vary considerably. For example, their moisture requirements differ widely. Plants, such as barley, maize or sugar cane need well aerated soils for optimum growth. On the other hand, some rice plants require wet anaerobic conditions for the early part of their life cycle.

Soil fertility is usually discussed in the context of crop production but it can be considered from the point of view of inherent soil fertility and induced soil fertility. Nearly every soil has a certain inherent fertility. Soils that are wet, acid, alkaline or deficient in a particular element will support a specific plant community. Therefore they can be regarded as fertile with regard to the plants growing on them, but when humans want to replace one of these natural plant communities with a crop, the inherent fertility may not be suited to the particular crop. Then it may be necessary to change the soil to induce the type of fertility to suit the needs of the crop. Nevertheless, it is possible to set out in general terms the plant-soil requirements and to state the factors affecting these requirements.

Factors affecting plant growth:

1. Root-room and root-hold
2. Aeration
3. Moisture
4. Temperature
5. Essential elements
6. pH
7. Stable site

1. Root-room and root-hold

Most plants extend their roots into the soil for anchorage as well as to extract nutrients and water. Although the greatest amounts of roots occur within the top 15 cm of the soil some can penetrate to depths of over 4 metres. Therefore the thickness of soil available for root penetration is very important. Soils may be shallow because of rock or a compacted horizon near to the surface or due to waterlogging caused by a high water-table. Whereas it is usually impossible to increase the thickness of a soil over rock, it is often possible to lower the water-table by drainage, using the techniques mentioned below. Thus drainage increases the volume of soil available to the plants. However, there is a limit to such an increase due to difficulties in removing the excess water (Fig. 6.1).

With crops such as barley, maize and sugar cane, root-room restricted by rock usually means a reduction in the potential nutrient supply and available moisture. On the other hand with tree crops there is an additional factor of poor stability due to shallow rooting, thus trees grown on such soils are very liable to be blown over.

Drainage

The drainage of a soil can be achieved by a number of relatively simple techniques. In most temperate countries mole drainage or tile drainage is used. Mole drainage is carried out by drawing through the soil at a given depth a bullet-shaped object (the "mole") which forms a continuous passage into which water can percolate and move. The passages are laid out to connect with tile drains which carry the water away. Tile drainage involves digging a trench and laying short lengths of clay pipe end to end and then filling the trench. This is a more permanent and effective method than mole drainage and is the main method used in Britain. In a number of cases mole drainage is used to supplement tile drainage. At present, tests are being conducted to evaluate continuous lengths of perforated plastic pipes which are laid by a machine that draws the pipe into the soil without having to dig a trench. If the tests are successful a considerable saving in the cost of drainage could result.

In countries with very heavy rainfall and much run-off it is customary to use open ditches. These are extremely effective, but they need a considerable amount of maintenance. Further, they can induce erosion if they are not well laid out.

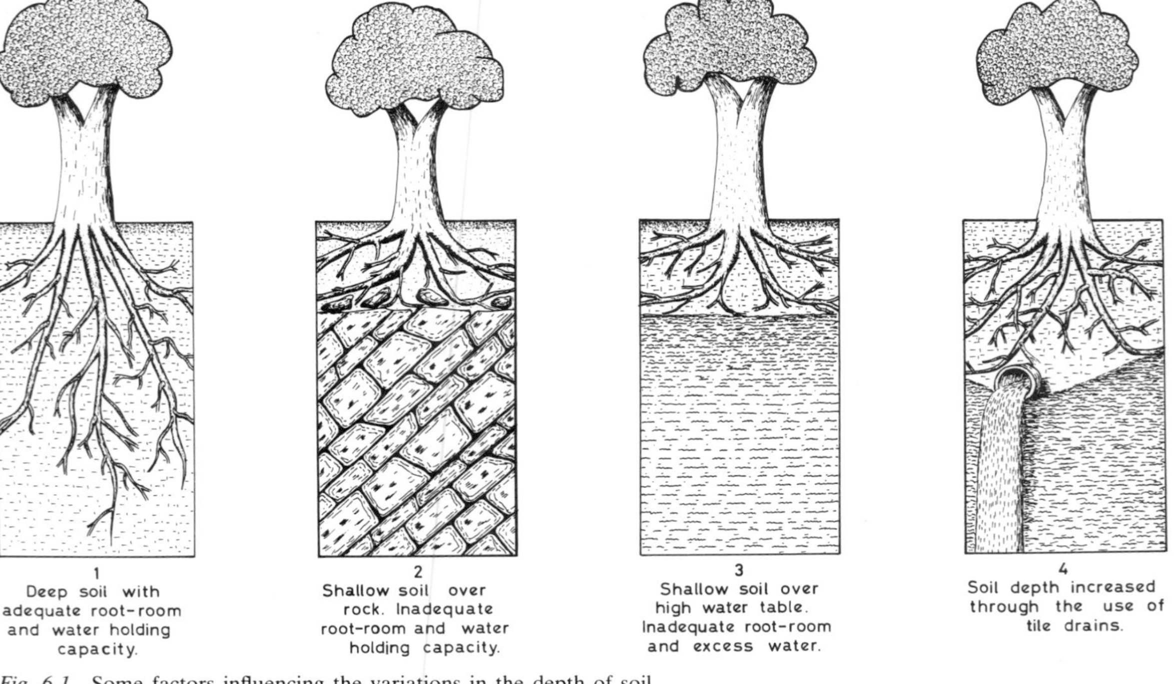

1
Deep soil with adequate root-room and water holding capacity.
2
Shallow soil over rock. Inadequate root-room and water holding capacity.
3
Shallow soil over high water table. Inadequate root-room and excess water.
4
Soil depth increased through the use of tile drains.

Fig. 6.1 Some factors influencing the variations in the depth of soil

In some cases toxins are produced under anaerobic conditions. Ethylene is produced in wet soils and inhibits root growth while cyanide is formed in the roots of peach trees causing their death (Greenwood 1970).

Drainage has a number of beneficial effects on the soil:

1. Increases the temperature in cool countries
2. Improves the structure
3. Allows air to enter
4. Increases the rate of organic matter decomposition
5. Improves the bacterial population
6. Improves germination
7. Allows a greater variety of crops to be grown
8. Reduces the incidence of plant disease such as blight
9. Causes fertilizers to be more effective and not washed out of the soil
10. Improves the health of livestock by reducing the incidence of foot rot and the frequency of the snail that is the intermediate host of liver fluke in sheep.

Perhaps it should be pointed out that while drainage is often beneficial to agriculture it may be at times an ecological hazard by threatening rare plant communities. For example, the drainage of dune slacks will threaten species such as *Baldellia ranunculoides* and *Epipactis palustris*.

2. Aeration

Plant roots and soil organisms require a constant supply of oxygen for respiration. Those plants such as rice that normally root in anaerobic conditions have a special mechanism for transferring oxygen from the stem and leaves to the roots. Since most roots and organisms develop in aerobic conditions they derive their oxygen from the soil atmosphere which usually contains an adequate supply (see page 104). A restricted or reduced supply of oxygen can be due to soil compaction, a high water-table or to the accumulation of carbon dioxide due to a lack of interchange between the soil and above ground atmosphere. Good aeration is usually facilitated by a granular or crumb structure and free drainage.

3. Moisture

An adequate and balanced supply of moisture is essential for plant growth. Moisture is constantly being taken up by plants together with nutrients and is lost by transpiration. It is estimated that 1 kg of dry weight increase in plants requires about 500 kg of transpired water. Thus a grain crop yielding 10 t ha^{-1} will transpire 2000–5000 tonnes which is equivalent to 200–500 mm of rainfall, therefore a steady supply is necessary if growing plants are to remain alive. Under certain extreme conditions plants may lose more water than they take up even though there may be an adequate supply in the soil. This condition is known as physiological drought and occurs commonly during the day in very hot climates but the plants recover during the cool of the night.

The moisture in soils can be considered in terms of input, retention and losses.

Fig. 6.2 Furrow irrigation prior to cropping

Moisture input

The moisture entering the soil is derived from three main sources, rainfall, melting snow and irrigation. In humid climates the input by rainfall or from melting snow is usually adequate, but in arid and semi-arid areas an adequate system of agriculture can be sustained only by irrigation which may take many forms. The water may be run on to a flat land surface causing complete flooding for a given period and then the excess water drained away. Alternatively the water is run into furrows if the plant roots are likely to suffer from oxygen deficiency as a result of complete flooding and waterlogging (Fig. 6.2).

Sprinkler or overhead irrigation is often practised on small areas or when the ground is sloping and therefore unsuited to flooding. This system is usually portable and can be relatively inexpensive reducing the need to install and maintain ditches to conduct the water (Fig. 6.3). But these systems are wasteful of water because of high evaporation. Tube irrigation to individual plants is now being tried and has many advantages; water wastage is reduced to a minimum and in addition fertilizers can be added to the water as required.

Moisture retention

The moisture retained in the soil will depend upon the amount removed and the speed of removal.

Water will percolate rapidly through the soil if it is very porous, through being very sandy or because of a well developed

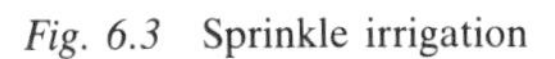

Fig. 6.3 Sprinkle irrigation

structure, thus the retention is likely to be very low. Fine textured and organic soils have smaller pore spaces and the particles themselves can absorb moisture, therefore moisture retention is higher and moisture movement is slower.

Thus, texture, organic matter content and structure affect the movement and retention of moisture in soils. Generally clays and organic soils have the highest moisture retaining capacity, while silts and organic soils have the highest available moisture. Clays retain more water than silts but a higher proportion is strongly held and therefore unavailable to plants. Where water is a limiting factor moisture conservation can be achieved by mulching, contour ploughing, dry farming and snow traps.

Moisture losses

The moisture retained in the soil is lost mainly by evapotranspiration. Therefore the rate of loss will depend upon temperature and plant cover, so that as temperature and plant cover increase, moisture losses will also increase. However, only part of the capillary water retained in the soil is available to be taken up by plants which will wilt and die after the available moisture has been exhausted.

It is probably correct to say that on the world scale, water is the main limiting factor to plant growth, for even in humid areas such as the east of England, supplementary irrigation in most years can accomplish a substantial increase in crop production. In many semi-arid areas where irrigation is not possible various methods of moisture conservation have to be practised such as dry-farming (see Glossary page 230). Thus for sustained agriculture, water management is an essential requirement. Generally the farmer has no control over the rainfall but management can reduce a deficiency or an excess.

4. Temperature

Most plants have their optimum growth within a specific temperature range. In cool climates germination will not start until the temperature is about 5°C. This restriction becomes increasingly important towards the poles with their long winters and cool spring weather.

On the other hand, high soil temperatures may result in an excessive loss of moisture with drought as a consequence.

5. Essential elements

There are a number of elements essential for both plants and animals. These can be divided roughly into the macroelements and the microelements. The macroelements are required in relatively large amounts whereas the microelements are required in only small amounts. Furthermore, these elements must be present in the correct proportions for when there is a deficiency or excess of any one element plant growth can be affected seriously and the plants develop symptoms of nutrient starvation or toxicity. Since animals and humans live directly or indirectly on plants, sometimes they can develop deficiency or toxicity as a result of plants being affected.

The 17 elements essential for plant growth are:

Macroelements	*Microelements*
Carbon	Manganese
Hydrogen	Copper
Oxygen	Zinc
Nitrogen	Molybdenum
Phosphorus	Boron
Potassium	Chlorine
Calcium	Iron
Magnesium	Cobalt
Sulphur	

Most of the elements are derived initially from the weathering of minerals and taken up by the plant roots. They are then returned to the soil in the plant litter which decomposes to release the elements which are again taken up by the plants. This cyclic process is fundamental in certain natural plant communities which are sustained purely by this process, excellent examples being some rain-forests.

The functions of most of the elements are not fully known but they affect plants in the following manner:

1. They form plant tissue
2. They act as catalysts and intermediates in a wide range of metabolic processes.

Each element plays a specific role as follows:

Carbon, hydrogen and **oxygen** are the major constituents of plant tissue and are derived from the atmosphere and water.

Nitrogen is derived from the atmosphere or dead tissues and in both cases it is transformed by bacteria in the soil into ammonia and nitrate which are taken up by the plant roots (see page 68). It occurs in great quantities in young plants particularly the leaves. Nitrogen forms a part of every living cell, occurring in chlorophyll and all proteins with many of the latter serving as enzymes. Abundance of nitrogen leads to green succulent growth while nitrogen deficiency causes a loss of colour, reduction in protein production and a gradual yellowing and stunted growth.

Deficiencies can occur in all soils due to the leaching of ammonia and nitrate and denitrification in wet soils. The presence of large amounts of fresh plant residues causes bacteria to compete with the plants for the available nitrogen. Deficiencies are easily rectified by the addition of fertilizers but excessive fertilization should be stringently avoided because much of the excess will ultimately cause pollution of the natural waterways.

Phosphorus is a constituent of every living cell and occurs in the protoplasm, with its greatest concentration in seeds thereby increasing their production. Phosphorus deficiency causes a purplish colouration at the seedling stage with later yellowing, stunted growth and delayed maturity.

Fixation by iron, aluminium and calcium causes deficiencies. Raising the pH to 5.5–6.5 reduces the deficiency but a complete cure can only be achieved by adding fertilizers. Even so, only about 25 per cent of the phosphorus added is taken up; the remainder is fixed.

Potassium is essential in all cell metabolic processes although the exact nature is not known. Apparently, it influences the uptake of other elements and affects both respiration and transpiration. It also encourages the synthesis and translocation of carbohydrates thereby encouraging cell wall thickening and stem strength. A deficiency can cause lodging in cereals and yellowing of the leaf tips and margins and is usually the result of leaching and continuous cultivation. This can be rectified by fertilizer application.

Calcium in the form of calcium pectate forms part of the cell wall structure and is necessary for the growth of the meristem (see Glossary, page 236). A deficiency leads to malformation of the growing parts but symptoms are seldom seen in the field.

Plants obtain their calcium from the soil minerals or from lime added to the soil.

Magnesium is active in enzyme systems and forms part of the chlorophyll. A deficiency causes discolouration and sometimes, premature defoliation of the plants. Animals eating such vegetation develop hypomagnesemia. A cure is achieved by adding a magnesium fertilizer or liming with magnesium limestone.

Sulphur occurs in some amino acids and also in the oils of some plants such as cabbages and turnips. A deficiency leads to stunting and yellowing but can be cured by the addition of gypsum. Deficiencies are becoming more common by a change in fertilizer practice away from such fertilizers as ammonium sulphate and also reduced atmospheric pollution.

Iron and **manganese** play a role in enzyme systems and are necessary for the synthesis of chlorophyll. The activity of these two elements is interrelated since iron can be inactivated by an excess of manganese. Iron deficiency only becomes evident in younger leaves and is seen as a yellowing particularly in the intervein areas. This is known as *iron-chlorosis* and is seen most commonly on calcareous or alkaline soils due to the lack of chlorophyll. Alleviation on calcareous soils is very difficult but can be achieved by spraying with a solution of an iron chelate.

Manganese deficiency is similar to that of iron but the chlorosis is usually more marked with the whole of the intervein area losing its green colour. High pH and good drainage may cause deficiencies that can be cured by spraying the crop with a solution of manganese sulphate.

Boron appears to play a role in calcium utilisation and the development of the actively growing parts of the plant. A deficiency leads to such conditions as heart-rot of beet and internal cork of apples. It is also essential for the fixation of nitrogen by bacteria in the nodules of legumes. A deficiency is cured by applying borax in exactly the correct amount since the margin between deficiency and toxicity is very small. Toxicity prevents pigment production.

Copper and **zinc** form part of the enzyme systems and are necessary for the formation of growth promoting substances. Copper deficiency is common on peat leading to growth abnormalities such as rapid wilting and weak stalks, spiralling of leaves and no grain formation. For rapid results the crop should be sprayed with a solution of copper oxychloride, otherwise the soil should be sprayed with a copper sulphate solution. Zinc deficiency symptoms vary from plant to plant; mottled leaf of citrus is a well known example.

Molybdenum is necessary for the reduction of nitrate in the plant, otherwise nitrate will accumulate and interfere with protein synthesis. Nitrogen fixation by legumes is also dependent upon molybdenum. A deficiency is caused by acidity and results in plant abnormalities such as the formation of narrow leaves curled around the midrib in cauliflowers and no development of the "curd". This is cured by adding the correct amount of sodium ammonium molybdate to the soil since an excess of molybdenum seems to induce copper deficiency in animals.

Chlorine is responsible for regulating the osmotic pressure and cation balance in plants. A deficiency gives a wilting appearance and "bronzing" of the leaf.

Cobalt has only recently been found to be necessary for the formation of nodules and nitrogen fixation in legumes. It is not required by non-leguminous plants but is taken up by all plants.

Availability of essential elements

Soils can be thought of as the reservoir of most of the essential elements for plant growth. Soils that have high contents of primary minerals such as feldspars, amphiboles and pyroxenes contain a large reservoir of elements but strongly weathered soils may have only a small reserve. Although the total reserve of elements is a factor in determining soil fertility, probably more important is the degree of availability of the elements since in a number of cases the elements such as calcium may be tightly locked up within the silicate structure or in undecomposed organic matter and be unavailable or only slowly available to plants. Therefore a clear distinction has to be made between the total and available elements.

Elements such as calcium and magnesium when present as carbonates are usually easily available because carbonates are readily soluble in the soil solution. Generally, the available cations occur as exchangeable cations on the surface of the clay particles from which they enter into solution or are taken up directly by plant roots. Thus, the higher the base saturation the more cations there are available. The anions essential for plants also vary very much in availability. Nitrogen usually occurs as a part of the tissue of organisms and is unavailable until it is transformed by microorganisms into ammonia and nitrate which are easily available. Phosphates which occur principally as calcium phosphates are very slowly soluble and thus not readily available. Therefore, in order to know whether a soil is fertile or not it is necessary to know the degree of availability of the essential elements.

Various laboratory methods have been devised for assessing the type and amount of available elements in the soil. In these methods the soil is usually treated with extractants such as sodium bicarbonate or ammonium acetate and the elements determined in the filtrate. Then, based upon the experience of the chemist it is possible to state which elements are deficient and also their degree of deficiency. This technique is very rapid, allowing several hundred samples to be assessed in a single day but the interpretation of the results is very subjective because it depends very largely upon the experience of the operators.

The best methods for assessing soil fertility are field experiments which are designed to determine which elements are deficient and also to assess the amount of each deficient element which has to be added to the soil. This method also takes into consideration the effects of variations in the environmental factors such as climate. For example, it may be necessary to determine the amount of nitrogen required by wheat. Then a number of plots are laid out and an adequate amount of the main nutrients are added except nitrogen. Each plot is then planted with wheat and at the same time nitrogen is added at four or five different levels starting with no nitrogen and increasing in stages up to excess.

After the crop has grown it is harvested and the grain production weighed. Then a graph is prepared by plotting the weight of grain against nitrogen added and from the graph it is possible to determine the amount of nitrogen required for optimum grain production. This method is more reliable, but it is time consuming, particularly with crops other than annuals.

When investigating the nutrient requirements of plant species it may not be possible to carry out field experiments so that the soil scientist and ecologist usually have to rely on pot experiments. This method of investigation is advantageous when it is necessary to exercise rigorous control over the environment and particularly when trials are being conducted on microelements. The technique involves the collection of soil samples which are dried, ground and sieved so as to remove the coarse particles and to make them homogeneous. The soil is then placed in pots and seeds planted. Varying amounts of nutrients can be added before or after planting depending upon the nature of the experiment. The plants are then usually grown in a controlled environment in a greenhouse.

Many pot experiments are conducted using pure quartz sand or some other pure medium, particularly when testing the effect of microelements, and therefore offer advantages over field trials. In addition there is a considerable saving in space allowing numerous experiments to be conducted at the same time.

A further point is that the nutritional requirements of plants vary widely from species to species. For example, root crops take about twice as much nitrogen, potassium and calcium as cereals thus the assessment of the availability of nutrients always has to be made with reference to a specific crop.

When crop cultivation is carried out, the reservoir of elements in the soil is either insufficient or their production by weathering and microbial processes is too slow and therefore it is usual to add fertilizers, compost or manures. These vary considerably in type, the precise one being determined to a certain extent by their availability and the nature of the crop. Set out below is a list of the common fertilizers and the elements that they supply.

Sulphate of ammonia	20.5% N
Urea	45% N
Ammonium nitrate	35% N
Rock phosphate	11–15% P
Super phosphate	7–8% P
Basic slag	2–8% P

Bone meal	7–13% P
Potassium chloride	39–42% K

Some fertilizers are more soluble in the soil solution than others, therefore some elements are more available than others, so that in a number of cases the total content of an element in a fertilizer may not represent the amount available. This is particularly true of the phosphorus bearing fertilizers which are slowly soluble. In many cases fertilizers are added as mixtures to give the required amounts of nutrients.

In many places green manuring is practised. This involves the growing of a seasonal crop, particularly a legume and ploughing it in at flowering. This increases the N and humus content, thereby increasing the water-holding capacity.

Microelement deficiencies and toxicities

It is not usual to add microelements to soils because either they are released in sufficient amounts by the weathering of minerals or they may occur in sufficient amounts in fertilizers as impurities, but there are a number of well known situations with microelement deficiencies. For example, most plants growing on peat soils are deficient in copper because of its slow release by plant decomposition.

When discussing the deficiency and toxicity symptoms of elements it is not sufficient to consider them only in relation to plants. It is necessary, also, to consider them in the wider context of animals and humans who depend directly or indirectly upon plants for food, since deficiencies or toxicities in plants can be passed on to animals and humans. Many gardens near tanneries have been manured for long periods with composted leather waste which contains a high content of copper and chromium. These elements have been taken up in large amounts by the plants and passed on to the people eating the vegetables. It is suggested that the higher incidence of stomach cancer in such areas is due to the excessive amounts of these elements.

Toxicities can arise in a number of other ways. A particularly good example is the smoke from aluminium factories containing a high content of fluoride which may be toxic to the surrounding vegetation. Another example is the abnormally high content of lead in hedgerow soils due to the various lead compounds in petrol engine exhaust gases.

Some elements such as cobalt and selenium required by animals and humans may not be necessary for plant growth. Nevertheless, it is essential for plants to take up these elements so that they can be passed on to the animals and man when they eat the plants (Fig. 6.4). Cobalt deficiency in cattle is common in many parts of the world, particularly good examples being found in Australia. In Scotland selenium deficiency causes muscular dystrophy in sheep.

6. pH

Whereas the pH values of soil are fairly variable (see page 108) the pH values of most cultivated soils range from 5.5–7.5 but for each crop there is a fairly narrow range. Crops such as tea grow best at pH values about 4.5, others such as wheat have an optimum range of 6–7.5. The pH of the soil is maintained at a suitable level by the addition of liming material which is usually a form of ground limestone.

Since the application of materials to the soil is quite expensive it is customary to add liming materials once every five to ten years. This is made possible by the fact that the liming material is lost relatively slowly by leaching from the soil because of its low solubility.

The degree of acidity of the soil affects a number of soil properties and processes, more especially the activity of the micro and mesoorganisms. Those that are particularly beneficial to crop production, such as earthworms and bacteria, prefer conditions about neutrality. In some cases acid soils encourage the growth of certain plant pathogens such as club root which can be controlled by liming.

Liming also supplies calcium and improves the solubility of phophorus both of which are important in tropical soils where the precipitation of Al is also important.

High acidity may lead to the following:

1. increasing amounts of manganese and aluminium in the soil solution, these can be taken up by plants and are toxic

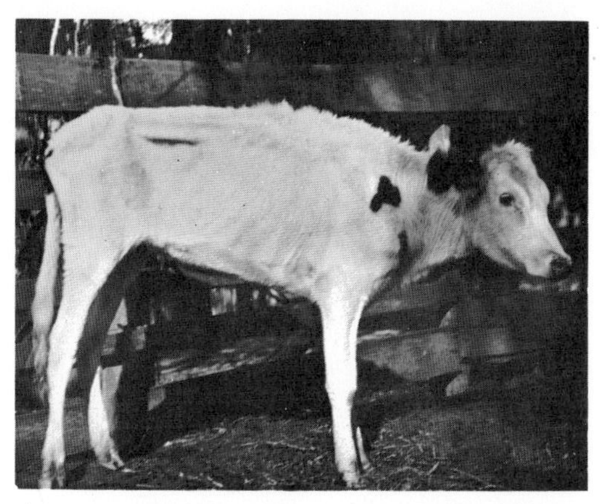

A

B

Fig. 6.4 (A) A calf with advanced symptoms of cobalt deficiency (B) The same animal on a balanced diet

2. little formation of ammonia or nitrate
3. low availability of phosphorus
4. molybdenum may be deficient
5. reduces root activity
6. certain disease such as club root thrive.

7. Stable site

In order to conduct a system of continuous crop production it is essential to have a stable site since continuous removal of material from the surface means the loss of the most fertile part of the soil as well as a steady and gradual reduction in soil thickness and root room. All sites are liable to erosion either by wind or water.

Wind erosion

Wind can remove particles of sand, silt and clay from any site and there are a number of very well known examples of catastrophic wind erosion such as the famous dust storms of the early 1930s in the midwestern prairie states of the USA. This erosion was a result of removing the natural grassy vegetation and leaving the soil surface exposed for long periods between crops (Fig. 6.5). A common form of wind erosion is moving sand dunes which are most frequent in coastal situations. These are a hazard because they can be blown onto adjacent agricultural land or across roads. Sometimes the volume of material that moves is sufficient to bury buildings and it is claimed that a complete village was buried at Culbin in north-east Scotland. More recently wind erosion in southern England is partly due to the removal of hedges.

Water erosion

It is on slopes that erosion by water is the greatest hazard and usually sites of more than a few degrees require some erosion control measures. There are three main types of water erosion, as determined by the volume and the speed of the water moving over the surface. Firstly there is *sheet erosion* (Fig. 6.6) which takes place when water moves evenly over the surface at a fairly slow speed. This form of erosion is insidious and can go on for many years without being noticed by anyone except the expert. As the speed of movement increases the water begins to cut into

Fig. 6.5 Top soil being blown off a bare surface

the surface of the soil causing the second type – *rill erosion*. The third type is *gully erosion* which is caused by a large volume of water moving rapidly over the surface, particularly on moderate or steep slopes (Figs. 6.7 and 6.8).

The material removed by sheet erosion often accumulates on the lower parts of slopes so that it is common to find the top soil increasing in thickness down the slope. The material removed by rill and gully erosion often enters streams and rivers where it can cause much damage and create many problems. For example, it can accumulate in reservoirs thereby reducing the water supply or it may accumulate in estuaries and harbours and be a hazard to shipping.

Fig. 6.6 Sheet erosion. The surface of the soil has been washed away exposing the roots which may eventually die due to desiccation

Rain splash has now been recognised as an important erosive agent for at least two reasons. When rain drops hit the surface of the bare soil they cause material to be splashed upwards and outwards. This is of little effect on flat surfaces but on slopes some of the soil is splashed down the slope. A further effect of rain drop impact is that it puddles and seals the surface thus reducing infiltration which leads to increased run-off and potential erosion.

Fig. 6.7 Gully erosion. This is the normal type of devastation that can result following the removal of vegetation

From the above it should be clear that soil erosion is bad because of the loss of valuable top soil and because of the problems it creates in reservoirs and the drainage system.

It should be mentioned also that soil is itself a reservoir for moisture. Some of the moisture from rainfall or melting snow is absorbed by the soil and percolates slowly through, eventually reaching streams and rivers, thus producing a fairly steady supply of moisture to the drainage. This is essential in order to maintain a constant domestic water supply, it also determines the nature of the fish-life in the river. If the moisture does not enter the soil but runs off over the surface after each shower of rain, then there will be wide fluctuations in the level of the rivers which could cause disastrous floods, leading to loss of life and property.

It is not possible to stop fluctuations in the level of most rivers,

Fig. 6.8 A deep gully that has undermined the tree, leaving the roots dangling in the air

but sometimes they can be reduced considerably if run-off is controlled. In contrast, flooding has always been extremely beneficial in some areas. The annual flood of the Nile is virtually the only source of water for crop growth. Here the flood waters are carefully controlled to ensure that there is maximum infiltration and storage of moisture for the complete life-cycle of the crops.

Erosion control

The amount of run-off and soil erosion can be reduced by a number of methods. Probably the simplest procedure is to keep the surface constantly covered by vegetation. However, this is not possible when arable cultivation is conducted but it should always be practised on steep slopes, by having forests, orchard crops or permanent grassland.

Wherever possible ploughing should be carried out along the contour so that there are no furrows down which water can run. A common form of erosion control is to construct terraces which restrict the speed of moisture movement as well as increasing greatly the infiltration on the flat terrace surface. The ancient terraces of many of the grape growing areas are world famous. This system not only reduces erosion but also stores moisture in the soil (Fig. 6.9). Another form of erosion control is strip cropping which consists of narrow strips of two or more crops grown alternately along the contour (Fig. 6.10). Usually each alternate strip is a biennial or perennial so that only alternate strips are bare at any one time, therefore any erosion that starts in the bare strip is checked by the crop on the next strip.

In many cases erosion is not confined to a single field or a single farm but affects a large area. Therefore control requires a communal effort and usually large capital input which for one reason or another may be unavailable.

Erosion is one of the greatest losses of soil fertility and could

Fig. 6.9 The terrace cultivation of rice in Java

Fig. 6.10 Contour strip cropping in Wisconsin, USA

have been a contributory factor towards the collapse of some ancient civilisations.

In many situations it is possible to predict the erosion potential of the soil when most of the soil and environmental factors are known. The factors are used by Wischmeier and Smith (1978) to produce a soil-loss equation.

Cultivation

With most systems of arable crop growth it is usual to cultivate the soil by the use of implements. The type and degree of cultivation will depend upon the type of crops that are grown as well as upon the economic status of the country. There are a number of reasons why the soil is cultivated. These include:

1. The production of an adequate depth of loose, porous top soil into which plants can freely extend their roots

2. The control of moisture, aeration and temperature
3. The destruction of pests and weeds
4. The mixing of plant remains and manures in the soil.

These ends can be achieved by a simple hand tool such as a hoe, by means of a primitive human or ox-drawn plough or by highly sophisticated tractor-drawn ploughs. At present about 70 per cent of all farming is done by hoes and wooden ploughs (Figs. 6.11 and 6.12). When growing grain crops such as wheat the soil is ploughed and then harrowed to produce an even surface and to break down the large clods. Seeds are then planted and at the same time fertilizer is added in such a manner that the concentration next to the seed is not too high, otherwise the germinating seedling may be killed or badly damaged.

If the crop is potatoes there may be further cultivation during the life-cycle of the plant. This will include weeding, ridging and a further application of fertilizer – a *top dressing*.

In addition to the factors discussed above there are a number of essential non-soil factors that plants require for adequate growth. These include solar radiation as a source of energy for photosynthesis, respiration and transpiration. Also required is

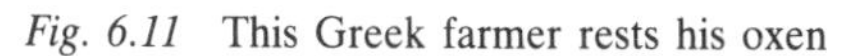

Fig. 6.11 This Greek farmer rests his oxen

Fig. 6.12 A modern tractor and plough in action

optimum atmospheric temperature which varies from species to species.

There has been a gradual move away from conventional tillage to systems of minimum cultivation and direct drilling, i.e. without ploughing and harrowing. This system relies heavily on modern herbicides to control weeds and sometimes to kill the previous crop prior to planting. Minimum cultivation reduces the likelihood of surface compaction and ploughpan formation and generally it is more effective on well drained soils.

Land use

Most human activities are concerned with the direct or indirect use of the land. As technology improves greater and greater demands are made upon the land so that more and more the potential of the soil is becoming a factor limiting personal well-being, since the soil is required to produce more food and to carry more buildings. Thus it is steadily becoming the focal point in conservation and environmental studies.

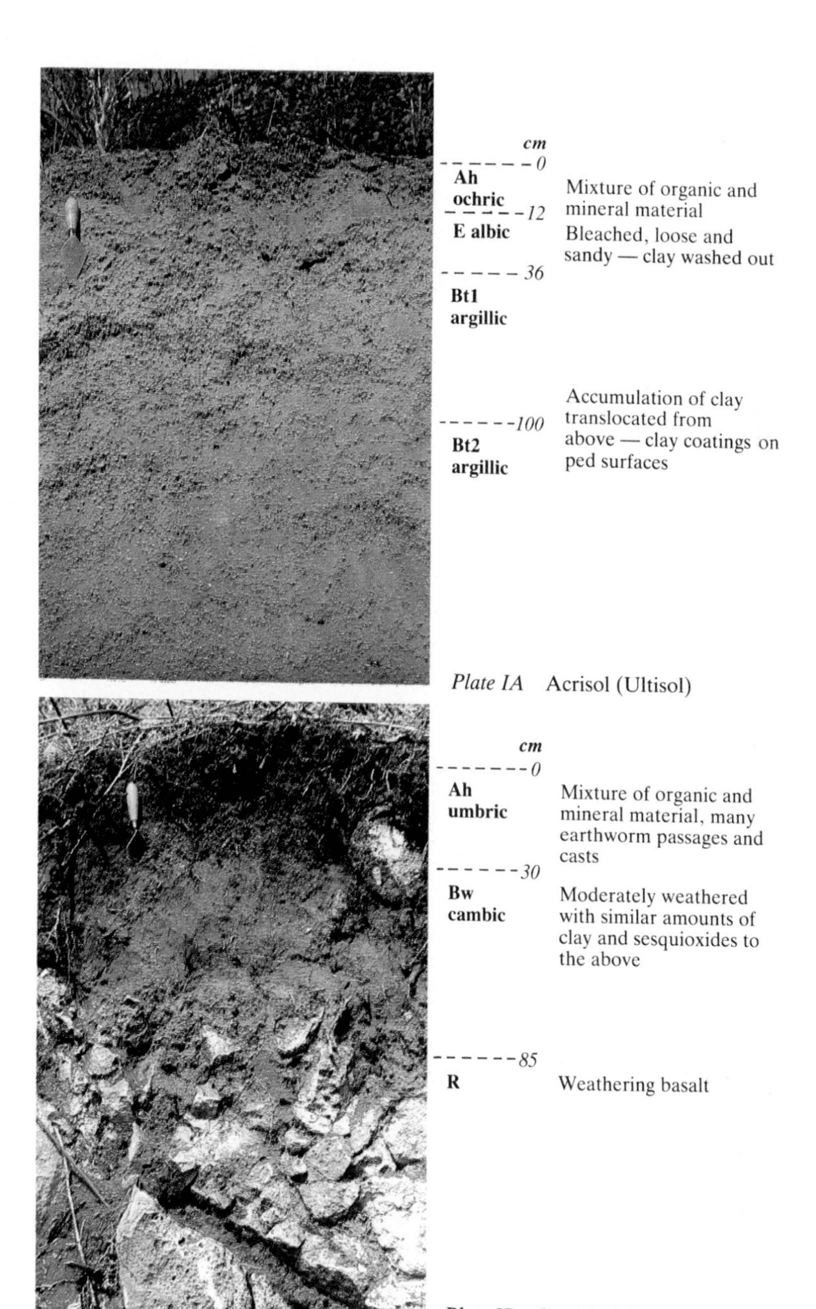

Plate IA Acrisol (Ultisol)

Plate IB Cambisol (Inceptisol)

cm

0

Ah mollic — Mixture of organic and mineral material, abundant earthworm passages and casts, numerous passages caused by the blind mole rat, pseudomycelium of calcium carbonate at the bottom

85

Ah/Ck — Abundant recent and old crotovinas, abundant pseudomycelium and concretions of calcium carbonate

150

C — Relatively unaltered loess with vertical earthworm passages

Plate IIA Chernozem (Mollisol)

Rainforest

Cultivated grassland

cm

0

Ah ochric ← Mineral material with a little organic matter

10

Bws oxic — Strongly weathered clay with well developed angular blocky structure, remains uniform for 3–4 metres then grades gradually into weathered rock underlain by solid rock

Plate IIB Ferralsol (Oxisol)

Whereas early humans cleared small areas of ground to create villages and to cultivate a few crops, modern people need the ground for:

1. Growth of crops and animals for food on a permanent basis or as shifting cultivation
2. Building sites for factories and houses
3. Grazing the natural vegetation
4. Building roads and airfields. It is estimated that in Britain 20,000 hectares per annum are used for roads
5. Construction of playing fields for football, golf, horseracing, etc.
6. Gravel pits and open-cast mining for coal, iron ore, etc.
7. Nature reserves and parks
8. Sewage disposal particularly the use of septic tanks in most continental and inland areas
9. Reservoirs, the water being used for power supply, domestic and industrial consumption and recreation.

Running in parallel with the growth in technology there has been an enormous growth in the world's population so that competition for the land has increased as a natural consequence. The nature of the competition is extremely complex, including rivalry between individuals for a given piece of land and the competition between humans and the other creatures in nature. Thus we find around towns that there is competition for the use of land for factories, crops, housing or playing fields to mention only a few uses. Competition is not confined to the areas peripheral to towns, it occurs also in the heart of rural areas as in the uplands of Britain where there is strong competition between forestry, sheep grazing and the shooting of deer and grouse. Throughout large parts of Australia one sees the very strong competition between sheep and the kangaroo for the relatively small amount of natural herbage. These various types of competition often lead to many types of misuse and damage to the land including the following:

1. Soil erosion due to deforestation, poor cultivation or death of vegetation caused by fumes from factories
2. Pest infestations such as potato eelworm due to continuous growth of potatoes
3. Pollution due to the addition of excess pesticides and herbicides

4. Pollution of the surface soil due to poor siting or over-loading of septic tanks. This can cause the spread of hepatitis
5. Salinisation due to poor cultivation and irrigation with saline water
6. Dumping of industrial waste creating slag heaps and the like.

Although the nature of the soil itself should be one of the major factors determining its use, this may not be the case when economic considerations come first, often leading to misuse of the soil. Thus it may be possible to produce a land capability classification for a given area but implementing it might not be possible because of some specific local custom or the lack of capital. Probably taking the world as a whole it is ignorance or a low level of technology that is often responsible for the misuse or improper use of land. It is in countries with a high level of technology that one tends to find the best use of land and the highest level of production but there are notable exceptions to this in the form of peasant agriculture in many parts of Western Europe and catastrophic erosion in the USA and Australia. Thus at present we find similar soils being used in a variety of ways. In one area a given soil may have a well developed and balanced system of land use which has been built up following a soil survey. This would include agriculture, forestry and wild-life preservation, but in another area a similar soil may be supporting subsistence agriculture because of the lack of capital. In some countries efforts are being made to make good some of the damage that has been done, but in other cases it is too late, as in central Turkey where many hillsides have been eroded down to the bare rock. An excellent example of reclamation is in the Tennessee Valley in the USA, where, by building dams and hydro-electric stations across the river, enough water and electric power are produced to change the standard of agriculture from subsistence to prosperity. This scheme plus others have shown that prosperity comes only when the utilisation of an area takes into consideration all the facets of the environment such as:

1. Soil erosion control
2. Maintenance of soil fertility by the addition of lime, fertilizers, drainage, etc.
3. Forest management and afforestation when needed
4. Wild life preservation

5. Elimination of pollution of the atmosphere and natural waters.

Catchment studies

Soils studies now usually form an integral part of comprehensive environmental investigations particularly in catchment studies. These are concerned with such aspects as the total amount of rainfall in a catchment, its utilisation by the vegetation and monitoring its movement over and through the soil, thereby enabling prediction of erosion potentials and the gains and losses of ions in the system. The gains are from the atmosphere often through acid rain pollution and the losses are by leaching and run-off. Sometimes these losses are in solution but in cases of rapid run-off they may be as fine particles in suspension. Catchments can range from undisturbed natural sites to those that are completely cultivated. The monitoring of cultivated catchments is of great importance in determining the amount of excess fertilizers, herbicides and pesticides entering the drainage which could develop unacceptable levels of pollution. Thus catchment studies have the potential to predict the best types of land use.

7 World Soils

The soil units are those used in the legend of the FAO-Unesco soil map of the world. They do not correspond to equivalent categories in different classification systems but they are generally comparable to the "great group" level.

An attempt has been made to use as many traditional names as possible, such as Chernozems, Kastanozems, Podzols, Planosols, Solonetz, Solonchaks, Rendzinas, Regosols and Lithosols. Names which in recent years have acquired a more general acceptance like Vertisols, Histosols, Rankers, Andosols, Gleysols and Ferralsols have also been adopted. For a limited number of soils it was necessary to coin new words. These include Luvisols, Acrisols, Cambisols, Phaeozems, Yermosols, Xerosols, Nitosols and Arenosols. The names of the individual units are achieved by adding an adjectival prefix to the main classes such as Orthic Acrisols, Cambisols, Phaeozems, Yermosols, Xerosols, Nitosols

Acric from Gr. *akros* = ultimate; connotative of very strong weathering and having a CEC of 15 meq kg^{-1} clay or less.

Albic from L. *albus* = white; connotative of strong bleaching.

Calcaric from L. *calcium*; connotative of the presence of calcium carbonate by accumulation or being present originally in the parent material.

Calcic from L. *calxis* = lime; connotative of strong accumulation of calcium carbonate to form a calcic horizon and sometimes an accumulation of gypsum to form a gypsic horizon.

Dystric from Gr. *dys* = ill, dystrophic, infertile; having a base saturation of <50%. For Histic H horizons having a pH of <5.5 in at least a part.

Cambic from L. *cambiare* = change; connotative of changes in colour, structure, or consistence resulting from weathering *in situ*.

Chromic from Gr. *chromos* = colour; connotative of soils with a high chroma.

Eutric from Gr. *eu* = good; eutrophic, fertile, having a base saturation of >50 per cent. For Histic H horizons having a pH of 5.5 or more throughout.

Ferralic from L. *ferrum* and *aluminium*; connotative of a high content of sesquioxides and CEC of <240 meq kg^{-1} clay.

Ferric from L. *ferrum* = iron; connotative of "ferruginous" soil and having ferric properties or high iron in the B horizons of Podzols.

Gelic from L. *gelu* = frost; connotative of permafrost.

Gleyic from Russian local name *gley* = mucky soil mass; connotative of an excess of water.

Glossic from Gr. *glossa* = tongue; connotative of tonguing of the A horizon into underlying layers.

Gypsic from L. *gypsum* = gypsum; connotative of strong accumulation of gypsum and the formation of a gypsic horizon.

Haplic from Gr. *haplos* = simple; connotative of soils with a simple, normal horizon sequence.

Humic from L. *humus* = earth; rich in organic matter and usually an umbric A horizon and/or humus in the middle horizon.

Leptic from Gr. *leptos* = shallow; connotative of weak development.

Luvic from L. *luvi* = to wash; connotative of illuvial accumulation of clay and often an argillic B horizon.

Mollic from L. *mollis* = soft; connotative of good surface structure and a mollic A horizon except in the case of Mollic Gleysols.

Orthic from Gr. *orthos* = true; connotative of common organic matter.

Orthic from Gr. *orthos* = true; connotative of common occurrence.

Pellic from Gr. *pellos* = dusky, lacking colour; connotative of soils with a low chroma.

Placic from Gr. *plax* = flat stone; connotative of the presence of a thin iron pan.

Plinthic from Gr. *plinthos* = brick; connotative of mottled clayey materials which harden irreversibly upon exposure.

Rhodic from Gr. *rhodon* = rose.

Solodic from Russian *sol* = salt; connotative of soils high in salt.

Takyric from Uzbek *takyr* = barren plain.

Thionic from Gr. *theion* = denoting the presence of sulphur; having a sulphuric horizon or sulphidic material or both.

Vertic from L. *verto* = turn; connotative of turnover of the surface and having vertic properties.

Vitric from L. *vitrum* = glass; connotative of soils rich in vitric material.

Xanthic from Gr. *xanthos* = yellow; connotative of soils having a yellow colour.

Acrisols (Colour Plate IA)

APPROXIMATE EQUIVALENT NAMES: Red-yellow Podozolic soils; Ultisols (USDA); Yeltozems (USSR); Luvosols (EAF).

GENERAL CHARACTERISTICS: These are the tropical and subtropical soils of old landscapes that have a monsoon climate and are extremely weathered and leached. They have a red, brown or yellow argillic B horizon with a base saturation of less than 50 per cent. They usually have an ochric or umbric A horizon and one or more of the following: a high content of organic matter in the B horizon, ferric properties, plinthite or hydromorphic properties. These soils are generally of low fertility because of both macro and micronutrient deficiencies often coupled with aluminium toxicity. Additions of lime and fertilizers are essential but this often proves to be too expensive. In addition they are highly susceptible to erosion if used for arable cultivation. Therefore they are best reserved for forestry and grazing.

SUBDIVISIONS: *Orthic Acrisols* have an ochric A horizon and an argillic B horizon. *Ferric Acrisols* have ferric properties – coarse red mottles or nodules up to 2 cm in diameter or a low CEC (<40 meq kg^{-1} clay). *Humic Acrisols* have an umbric A horizon or high content of organic matter in the B horizon. *Plinthic Acrisols* have plinthite within 12 cm of the surface. *Gleyic Acrisols* have hydromorphic properties within 50 cm of the surface.

Andosols

APPROXIMATE EQUIVALENT NAMES: Yellow brown loams and yellow brown pumice soils (New Zealand); Kuroboku (Japan); Andepts (USDA); Volcanic soils (USSR).

GENERAL CHARACTERISTICS: These soils are developed in volcanic ash and generally have a dark mollic or umbric A horizon overlying a brown cambic B horizon and one or both of the following: bulk density of less than 0.85 g cm^{-3}; 60 per cent or more vitric volcanic ash. These soils are distinguished by a high content of allophane which causes the low bulk density and gives a smeary consistence – thixotropic.

The natural fertility of these soils is high when formed in basic ash but many have low fertility mainly because of their great capacity to fix phosphorus, therefore it is advantageous to plant crops such as sweet potatoes (*Ipomea batatas*) that seem to show little response to the addition of phosphorus.

SUBDIVISIONS: *Ochric Andosols* have an ochric A horizon and a cambic B horizon; *Mollic Andosols* have a mollic A horizon; *Humic Andosols* have an umbric A horizon; *Vitric Andosols* lack a smeary consistence.

Arenosols

APPROXIMATE EQUIVALENT NAMES: Red and yellow sands, Psamments (USDA).

GENERAL CHARACTERISTICS: These soils are developed from coarse-textured unconsolidated material except recent alluvium. They have no diagnostic horizons other than an ochric A horizon but may consist of albic material more than 50 cm deep or may show characteristics of argillic, cambic or oxic B horizons. These however do not qualify as diagnostic horizons because of their coarse texture. These soils generally have a very low natural fertility but sometimes can give good crop yields when they occur in humid areas and supplied with fertilizers. When exposed they erode very rapidly especially by wind which will cause the formation of sand dunes that can overwhelm villages as is occurring in parts of Africa.

SUBDIVISIONS: *Cambic Arenosols* have cambic like features; *Luvic Arenosols* have lamellae of clay accumulations; *Ferralic Arenosols* have ferralic properties, i.e. a cation-exchange capacity of less than 240 meq kg^{-1}. *Albic Arensols* have albic material at least 50 cm thick.

Cambisols (Colour Plate IB)

APPROXIMATE EQUIVALENT NAMES: Brown forest soils, Brown Earths (General); Brunisols (Canada); Ochrepts, Umbrepts and Tropepts (USDA); Cinnamonic soils (USSR); Altosols (EAF).

GENERAL CHARACTERISTICS: This is a very widespread group of soils most of which show early stages of soil development. They have a cambic B horizon and may have an ochric or umbric A horizon, a calcic or gypsic horizon or a fragipan. The cambic B horizon may be absent when an umbric A horizon is present and thicker than 25 cm.

The ochric or umbric A horizon usually has a high content of earthworm vermiforms and other faunal faecal material.

Some Cambisols of the tropics show a considerable degree of weathering and are very close to the Acrisols or Ferralsols.

The relatively unweathered Cambisols of the cooler regions have a very high inherent fertility and usually carry a deciduous forest. When the forest is removed they can be adapted to a variety of systems of agriculture; more usually it is to mixed farming but large areas are used for dairying, orchards and other systems of land use.

SUBDIVISIONS: *Eutric Cambisols* have an ochric A horizon and a base saturation of 50 per cent or more between 20 and 50 cm from the surface. *Dystric Cambisols* have an ochric A horizon and a base saturation of less than 50 per cent between 20 and 50 cm of the surface. *Humic Cambisols* have an umbric A horizon which is thicker than 25 cm when a cambic B horizon is absent. *Gleyic Cambisols* have an ochric or umbric A horizon and show hydromorphic properties between 50 and 100 cm from the surface. *Gelic Cambisols* have permafrost within 200 cm of the surface. *Calcic Cambisols* have an ochric A horizon, a cambic B horizon and one or more of the following: a calcic horizon, a gypsic horizon or soft powdery lime within 125 cm of the surface, calcareous between 20 and 50 cm. *Chromic Cambisols* have an ochric A horizon and a base saturation of 50 per cent or more between 20 and 50 cm from the surface, have a strong brown to red cambic B horizon. *Vertic Cambisols* have an ochric A horizon and a cambic B horizon showing vertic properties. *Ferralic Cambisols* have an ochric A horizon and a cambic B horizon showing ferralic properties i.e. having a cation-exchange capacity of less than 240 meq kg^{-1} clay.

Chernozems (Colour Plate IIA)

APPROXIMATE EQUIVALENT NAMES: Black soils; Borolls (USDA); Chernozems (USSR).

GENERAL CHARACTERISTICS: These are the dark coloured soils of the steppe and some prairie grassland areas. They have a thick mollic A horizon and one or more of the following: a calcic or gypsic horizon, concentrations of soft powdery lime within 125 cm of the surface, an argillic B horizon.

The mollic A horizon is a tangled mass of faecal material of earthworms and/or enchytraeid worms. Most Chernozems, particularly those of Europe have passages (crotovinas) caused by burrowing vertebrates (Plate IIA). These together with the worms are responsible for the great thickness of the mollic A horizon.

These soils have a high nutrient status, excellent structure and high water-holding capacity which together impart a high natural fertility. Wheat, barley and maize are the principal crops but since these soils occur in low rainfall areas drought is a hazard.

SUBDIVISIONS: *Haplic Chernozems* have a thick mollic A horizon; *Calcic Chernozems* have a calcic or gypsic horizon; *Luvic Chernozems* have an argillic B horizon and *Glossic Chernozems* have a mollic A horizon that tongues into a cambic B horizon or the C horizon.

Ferralsols (Colour Plate IIB)

APPROXIMATE EQUIVALENT NAMES: Latosols, Lateritic soils, Ferralitic soils, (General); Oxisols (USDA); Krasnozems and Zheltozems (USSR).

GENERAL CHARACTERISTICS: These are the deep, red, brown and yellow, highly weathered soils of the humid tropics. They usually have an ochric A horizon overlying an oxic B horizon with a CEC <15 meq kg^{-1} clay and a very low base saturation. They tend to have moderate to high contents of clay, a well developed granular structure and a labyrinth of passages produced by termites. The clay mineralogy is dominated by kaolinite together with variable amounts of goethite, hematite and gibbsite. These soils are usually very deep, changing gradually through progressively less weathered material, to rotten rock then to the fresh rock (see page 77).

Because these soils are strongly weathered and occur under high rainfall conditions they are extremely nutrient deficient and acid, often with toxic levels of exchangeable aluminium. They also have a high capacity for fixing phosphorus. Thus lime, macro-nutrients and often micro-nutrients have to be added. The exchange capacity of the organic matter in these soils is very important because of the low CEC of the clay; however the organic matter is rapidly mineralised under the hot tropical conditions.

Although the soils are very poor for agriculture they carry a luxuriant natural forest sustained by nutrient cycling.

SUBDIVISIONS: *Orthic Ferralsols* have an ochric A horizon overlying a brown or reddish brown oxic B horizon. *Xanthic Ferralsols* have an ochric A horizon overlying a yellow oxic B horizon. *Rhodic Ferralsols* have an ochric A horizon overlying a red to dusky red oxic B horizon. *Humic Ferralsols* have an umbric A horizon or high organic matter in the oxic B horizon or both. *Acric Ferralsols* have an ochric A horizon and an oxic B horizon with CEC <15 meq kg^{-1} clay. *Plinthic Ferralsols* have plinthite within 125 cm of the surface.

Fluvisols

APPROXIMATE EQUIVALENT NAMES: Alluvial soils, Regosols; Fluvents (USA).

GENERAL CHARACTERISTICS: These are the very young soils of the recent alluvial deposits, deltas, estuaries and coastal situations. They have no diagnostic horizon other than an ochric or an umbric A horizon, a histic H horizon or a sulphuric horizon. They have a wide range of natural plant communities and occur in all parts of the world. They vary considerably in particle size distribution and are often stratified sometimes with buried surface horizons. Some Fluvisols are extremely fertile such as the alluvium of the Nile and many other large rivers.

SUBDIVISIONS: *Eutric Fluvisols* have a base saturation of more than 50 per cent between 20 and 50 cm from the surface. *Calcaric Fluvisols* are calcareous between 20 and 50 cm from the surface. *Dystric Fluvisols* have a base saturation of less than 50 per cent between 20 and 50 cm. *Thionic Fluvisols* have a sulphuric horizon or sulphidic material or both at less than 125 cm. If these soils are drained naturally or artificially the pyrite is oxidised to jarosite and sulphuric acid which makes them very acid. In addition the acidity dissolves aluminium which is toxic to plant growth. Thus these soils are very poor for crop growth.

Gleysols (Colour Plate IIIA)

APPROXIMATE EQUIVALENT NAMES: Gley soils; Aquents, Aquepts (USDA); Meadow soils (USSR); Subgleysols (EAF).

GENERAL CHARACTERISTICS: These are the wet mineral soils that are common in depressions in humid climates. They form from unconsolidated material exclusive of recent alluvium and show mottling and reduction within 50 cm of the surface. They have an ochric, mollic or umbric A horizon, a histic H, cambic B, calcic or gypsic horizon.

They are poorly drained soils but when drained they can be used very successfully for agriculture. In the tropics many Gleysols are used for growing rice (see below).

SUBDIVISIONS: *Eutric Gleysols* have a base saturation of >50 per cent between 20 and 50 cm. *Calcaric Gleysols* have a calcic or gypsic horizon within 125 cm and/or calcareous between 20 and 50 cm. *Dystric Gleysols* have a base saturation of <50 per cent. *Mollic Gleysols* have a mollic A horizon or a eutric histic H horizon. *Humic Gleysols* have an umbric A horizon or a dystric histic H horizon. *Plinthic Gleysols* have plinthite within 125 cm and *Gelic Gleysols* have permafrost within 200 cm.

An important group of Gleysols are the rice or paddy soils produced by cultivation and described below.

Rice soils

There are mainly two types of soils used for growing rice, one is a normal Gleysol with a high ground water table while the other is an aerobic soil that has been flooded repeatedly over many years. The morphology of the Gleysol is altered little by flooding but the morphology of the aerobic soil is changed markedly. The upper horizon develops grey reduced colours and reddish-brown iron pipes around the roots. Below is a horizon, mottled reddish-brown due to the translocation and deposition of iron. This is followed by a horizon with black manganese dioxide stainings. A property of both soils is the formation of a thin plough pan caused by ploughing and smearing the soil when it is wet.

Greyzems

APPROXIMATE EQUIVALENT NAMES: Argiborolls, Aquolls (USDA); Gray Forest soils (USSR).

GENERAL CHARACTERISTICS: These are the soils of cool continental areas, they developed beneath grassland bordering deciduous forest but are of limited distribution. They have a thick mollic A horizon with bleached coatings on ped surfaces. These soils have a high natural fertility and are used mainly for growing wheat, barley and maize.

Histosols

APPROXIMATE EQUIVALENT NAMES: Organic soils, Peat soils (General); Histosols (USDA); Bog soils (USSR).

GENERAL CHARACTERISTICS: These are the wet organic soils that have an H horizon of at least 40 cm. The organic matter accumulates because of the wet conditions and its composition is determined by the nature of the plant material. It varies from amorphous to fibrous and woody. Peat occurs mainly in the cool and cold temperate areas but also occurs in certain unique situations in the tropics especially in Sarawak.

Peat can be drained and used for producing very good crops but with time it is oxidised and shrinks. In some situations there are deficiencies of nutrients especially copper, cobalt, magnesium and boron.

SUBDIVISIONS: *Eutric Histosols* have pH values of 5.5 or greater; *Dystric Histosols* have pH values less than 5.5; *Gelic Histosols* have permafrost. Histosols may also be divided into Basin Histosols and Blanket Histosols. The former develop in depressions and vary from very acid to alkaline as determined by the nature of the water causing the wetness. Blanket Histosols form in upland cool humid areas and are always acid because they derive their water from rainfall and snow.

Kastanozems

APPROXIMATE EQUIVALENT NAMES: Ustolls (USDA); Chestnut soils (Canada); Kastanozems (USSR).

GENERAL CHARACTERISTICS: These are soils of the cool semi-arid areas with grassy vegetation. They have a brown to dark brown mollic A horizon at least 15 cm deep and one or more of the following: a calcic or gypsic horizon or soft powdery lime within 125 cm of the surface; some have an argillic B horizon.

These soils have a high inherent fertility and grain is the principal crop but the dry conditions impose severe utilization limitations.

SUBDIVISIONS: *Haplic Kastanozems* have a mollic A horizon; *Calcic Kastanozems* have a calcic or gypsic horizon; *Luvic Kastanozems* have an argillic B horizon.

Lithosols

APPROXIMATE EQUIVALENT NAMES: Lithosols (General); Lithic subgroups (USDA); shallow mountain soils (USSR).

GENERAL CHARACTERISTICS: These are the shallow soils of mountainous areas, fairly recent volcanic lava flows and areas scraped bare by ice. They are less than 10 cm deep and rest on hard rock. They have little potential for crop production but in humid areas they may produce enough vegetation for light grazing.

Luvisols (Colour Plate IIIB)

APPROXIMATE EQUIVALENT NAMES: Grey-Brown Podzolic soils; Alfisols (USDA); Sols lessivés (French); Cinnamonic soils (USSR).

GENERAL CHARACTERISTICS: This is a widespread group of soils that have an argillic B horizon with a base saturation >50 per cent. They may have an albic E horizon and one or more of the following: a calcic horizon, plinthite, vertic properties, ferric or hydromorphic properties.

Most Luvisols have a high potential for agriculture and although the base saturation is high most need liming and fertilizers in order to get maximum yields. The soils occur mainly in temperate regions where they form some of the major agricultural soils. They also occur in the tropics and subtropics where they are also agriculturally good soils.

SUBDIVISIONS: *Orthic Luvisols* have an argillic B horizon which is not strong brown or red and without an albic E horizon. *Chromic Luvisols* have a strong brown to red argillic B horizon and without an albic E horizon. *Calcic Luvisols* have an argillic B horizon and a calcic horizon. *Vertic Luvisols* have an argillic B horizon showing vertic properties. *Ferric Luvisols* have an argillic B horizon with ferric properties and no albic E horizon. *Albic Luvisols* have an albic E horizon overlying an argillic B horizon. *Gleyic Luvisols* have an argillic B horizon and hydromorphic properties. *Plinthic Luvisols* have an argillic B horizon and plinthite.

Nitosols

APPROXIMATE EQUIVALENT NAMES: Krasnozems (Australia); Terra Roxa (Brazil); Udalfs, Udults, Humults (USDA); Krasnozems (USSR).

GENERAL CHARACTERISTICS: These are clayey red soils of the tropics that have an argillic B horizon with shiny ped surfaces but without abrupt textural changes. Generally they have an ochric A horizon and varying base saturation. Some have an umbric A horizon and/or high organic matter in the B horizon. These soils are usually deep with a well formed sub-angular blocky or granular structure which imparts good root room and water storage. These are among the most fertile soils in the tropics and are extensively used for a wide range of crops. They do need fertilizers particularly phosphorus which is rapidly fixed.

SUBDIVISIONS: *Eutric Nitosols* have a base saturation of 50 per cent or more; *Dystric Nitosols* have a base saturation of less than 50 per cent and *Humic Nitosols* have a base saturation of less than 50 per cent and an umbric A horizon.

Phaeozems

APPROXIMATE EQUIVALENT NAMES: Brunizems; Tschernozems (Germany); Hapludolls (USDA); Chernozems (USSR).

GENERAL CHARACTERISTICS: These are the principal soils of the prairies of North America. They have a well developed dark coloured mollic A horizon and may have an argillic B horizon and hydromorphic properties. These soils have a very high inherent fertility and produce most of the grain grown in North America.

SUBDIVISIONS: *Haplic Phaeozems* only have a mollic A horizon; *Calcaric Phaeozems* are calcareous between 20 and 25 cm of the surface. *Luvic Phaeozems* have an argillic B horizon and *Gleyic Phaeozems* have an argillic B horizon and hydromorphic properties.

Planosols

APPROXIMATE EQUIVALENT NAMES: Pseudogleys, Stagnogleys (Germany); Aqualfs, Xeralfs, Ustalfs, Aquults (USDA); Podbels, Solods (USSR).

GENERAL CHARACTERISTICS: These are soils of flat continental areas with marked seasonality. They have an albic E horizon with hydromorphic properties and an abrupt junction to a slowly permeable underlying horizon. This may be an argillic or natric B horizon, a heavy clay pan or a fragipan. The slowly permeable horizon causes severe waterlogging during the wet season.

These soils are agriculturally very poor but they can produce reasonable yields of wheat in most years. They are also suitable for rice because of their slow permeability.

SUBDIVISIONS: *Eutric Planosols* have an ochric A horizon and 50 per cent or more base saturation; *Dystric Planosols* have an ochric A horizon and less than 50 per cent base saturation; *Mollic Planosols* have a mollic A horizon or a eutric histic H horizon; *Humic Planosols* have an umbric A horizon or a dystric histic H horizon; *Solodic Planosols* have more than 6 per cent sodium in the exchange complex in the middle horizon. *Gelic Planosols* have permafrost.

Podzols (Colour Plate IVA)

APPROXIMATE EQUIVALENT NAMES: Spodosols (USDA); Podzols (USSR).

GENERAL CHARACTERISTICS: These are the principal soils of the northern coniferous forest. They have a spodic B horizon and often a bleached albic E horizon. They may also have a thin iron pan or fragipan or hydromorphic properties (see pages 4 and 73). Podzols with very thick (>2 m) albic E horizons occur in many parts of the tropics where they develop in deposits of almost pure quartz sand. Podzols have an extremely low potential for agriculture since they need heavy applications of lime and fertilizers but they have proved to be productive in some parts of Europe and North America. They are often used for coniferous forestry or low volume grazing.

SUBDIVISIONS: *Orthic Podzols* have a spodic B horizon containing both iron and humus and have an albic E horizon. *Leptic Podzols* have a spodic B horizon containing both iron and humus but have no albic E horizon. *Ferric Podzols* have a spodic B horizon containing mainly iron and may have an albic E horizon; *Humic Podzols* have a spodic B horizon that contains mainly humus; *Placic Podzols* have an iron pan over the spodic B horizon; *Gleyic Podzols* have hydromorphic properties.

Podzoluvisols (Colour Plate IVB)

APPROXIMATE EQUIVALENT NAMES: Pseudogleys; Glossudalfs, Glossoboralfs (USDA); Derno-Podzolic soils (USSR).

GENERAL CHARACTERISTICS: These soils are common beneath the moist deciduous forests of the cool temperate areas. They are characterised by having an argillic B horizon with deep tonguing of overlying albic E horizon or iron concretions in the upper part of the argillic B horizon.

These soils have slow permeability which causes hydromorphism that reduces their potential for agriculture. With drainage and the application of fertilizers good crop yields can be obtained.

SUBDIVISIONS: *Eutric Podzoluvisols* have a base saturation of 50 per cent or more; *Dystric Podzoluvisols* have a base saturation of less than 50 per cent; *Gleyic Podzoluvisols* show hydromorphic properties within 50 cm of the surface.

Rankers

APPROXIMATE EQUIVALENT NAME: Lithic Haplumbrepts (USDA).

GENERAL CHARACTERISTICS: These are soils that have only an umbric A horizon less than 25 cm thick and developed on material other than alluvium; usually they are shallow due to the presence of rock. They have little agricultural value but in humid areas they may produce enough vegetation for light grazing.

Regosols

APPROXIMATE EQUIVALENT NAMES: Skeletal soils; Orthents, Psamments (USDA).

GENERAL CHARACTERISTICS: These are relatively recent soils developed on unconsolidated materials other than alluvium. They usually have an ochric A horizon and variable base saturation. Many have a potential for agriculture, the exact type being determined by local conditions. A significant use is the growth of coconuts on sands in the tropics.

SUBDIVISIONS: *Eutric Regosols* have a base saturation of 50 per cent or more; *Dystric Regosols* have a base saturation of less than 50 per cent; *Calcic Regosols* are calcareous; *Gelic Regosols* have permafrost.

Rendzinas

APPROXIMATE EQUIVALENT NAMES: Rendolls (USDA); Dern-carbonate soils (USSR).

GENERAL CHARACTERISTICS: These are the very dark coloured, shallow soils with a mollic A horizon and developed over calcareous material. These soils are very fertile but because of their shallowness they tend to have limited value for agriculture; however, a wide variety of crops can be grown in high rainfall areas.

Solonchaks

APPROXIMATE EQUIVALENT NAMES: Saline soils; Salorthids (USDA); Solonchaks (USSR).

GENERAL CHARACTERISTICS: These are the saline soils of the arid and semi-arid areas that contain appreciable amounts of soluble salts (see page 113). They often have an efflorescence of salt on the ground surface or on ped surfaces. Usually there are salt tolerant plants which may not form a complete cover. They may have a histic H horizon, a cambic B horizon, a calcic or gypsic horizon. Many solonchaks show hydromorphic properties and an associated high water-table.

These soils are of little value in their natural state but if the salts can be removed then they can be used for agriculture. This requires the soil to be washed with water but this is usually difficult because of the lack of salt-free water in these dry areas.

In many situations soils have been made saline by using poor irrigation water. This is a constant hazard in many areas where every year thousands of hectares of arable land are lost to cultivation by induced salinisation.

SUBDIVISIONS: *Orthic Solonchaks* have an ochric A horizon; *Mollic Solonchaks* have a mollic A horizon; *Takyric Solonchaks* have takyric features; *Gleyic Solonchaks* have hydromorphic features within 50 cm of the surface.

Solonetz

APPROXIMATE EQUIVALENT NAMES: Natrustalfs, Natrixeralfs, Natrargids, Nadurargids (USDA); Solonetz (USSR).

GENERAL CHARACTERISTICS: These are soils of semi-arid areas that have a natric B horizon, characterised by a sharp increase in clay, columnar or prismatic structure and high alkalinity. They may have an ochric or mollic A horizon, an albic E horizon and hydromorphic properties at depth. These soils are of little value for agriculture because of their alkalinity and high exchangeable sodium. Sometimes they can be improved by deep ploughing and the addition of gypsum. Following amelioration they give very good yields of wheat in Canada but in tropical countries they are not very productive.

SUBDIVISIONS: *Orthic Solonetz* have an ochric A horizon; *Mollic Solonetz* have a mollic A horizon and *Gleyic Solonetz* have hydromorphic properties. Similar soils that have a well developed albic E horizon that tongues into the natric B horizon are termed Solodic Planosols.

Vertisols

APPROXIMATE EQUIVALENT NAMES: Black earths; Regurs; Vertisols (USDA); Compact soils (USSR).

GENERAL CHARACTERISTICS: These are the dark coloured very clayey soils of many of the plains of the semi-arid and arid tropics and subtropics. They have 30 per cent or more clay usually dominated by montmorillonite which causes the soil to shrink and crack during the dry season and to swell during the wet season. The shrink–swell processes create pressures in the soil causing it to rupture and to form slickensides as one surface slides over another. It also causes the formation of a microtopography known as gilgai.

The surface of Vertisols may have a crust or a very well developed granular structure (Fig. 4.12).

Vertisols have a high potential for agriculture but some

inherent problems have to be overcome. The first is the low rainfall but this has been solved by irrigation in a number of the largest Vertisol areas particularly the Sudan where there is an abundance of water from the Nile. Secondly, because of the high clay content there is a narrow moisture range when the soils can be cultivated. When these problems are overcome productivity is very high.

SUBDIVISIONS: *Pellic Vertisols* are very dark and almost black, and become waterlogged during the rainy season. *Chromic Vertisols* are browner and tend not to be waterlogged.

Xerosols

APPROXIMATE EQUIVALENT NAMES: Aridisols (USDA); Semi-desert soils (USSR).

GENERAL CHARACTERISTICS: These are the soils of the semi-arid to arid areas that have a *weak* ochric A horizon and one or more of the following: a cambic B horizon, an argillic B horizon, a calcic horizon or a gypsic horizon. Often the calcic or gypsic horizon is very hard. They have an aridic soil moisture regime because precipitation is generally less than 200 mm per annum and usually falls as short intense showers. The natural vegetation is composed mainly of xerophytes thus normal crop production is impossible. However, they often have a very high natural fertility and produce good crops with irrigation. These areas are usually grazed but over-grazing can lead to the complete destruction of the natural vegetation, exposing the surface to erosion by both wind and water.

SUBDIVISIONS: *Haplic Xerosols* have a weak ochric A horizon and a cambic B horizon; *Calcic Xerosols* have a calcic horizon; *Gypsic Xerosols* have a gypsic horizon and *Luvic Xerosols* have an argillic B horizon.

Yermosols

APPROXIMATE EQUIVALENT NAMES: Aridisols (USDA); Desert soils (USSR).

GENERAL CHARACTERISTICS: These are the soils of the arid areas that have a *very weak* ochric A horizon and one or more of the following: a cambic B horizon, an argillic B horizon, a calcic horizon, a gypsic horizon or takyric features. They have a very sparse xerophytic vegetation and can only be lightly grazed. Like Xerosols, over-grazing destroys the vegetation and wind erosion follows. Many of these soils have a high inherent fertility and give goods crops when irrigated. Many of these soils are polygenetic and form in pre-weathered materials. In contrast some have developed from gypsiferous materials which are completely unsuited to crop growth even when irrigated.

SUBDIVISIONS: *Haplic Yermosols* have a *very weak* ochric A horizon and a cambic B horizon; *Calcic Yermosols* have a calcic horizon; *Gypsic Yermosols* have a gypsic horizon; *Luvic Yermosols* have an argillic B horizon and *Takyric Yermosols* have takyrs.

There are two other systems of soil classification that should be mentioned briefly, that of the USDA (1975) known as Soil Taxonomy and that of FitzPatrick (1980).

Soil Taxonomy

In Soil Taxonomy there are ten orders usually differentiated by the presence or absence of diagnostic horizons or features that show the dominant set of soil-forming processes. Thus they are created in a subjective manner. The orders are divided into suborders, great groups, subgroups, families and series. The names of the orders are all coined words with the common ending – *sol* such as *Spodosols*. A formative element is abstracted from the name and used as an ending for the names of all suborders, great groups and subgroups of one order. In the case of the Spodosols the formative element is – *od* –. Each suborder name consists of two syllables. The first indicates the property of the class and the second is the formative element of the order. Thus the *Orthods* are the common Spodosols. (Gk. *Orthos*. = true.) The names of the great groups are produced by adding one or more prefixes to the name of the suborder. Thus the *Cryorthods* are the cold Orthods (Gk. *Kuros* = cold). The subgroup names consist of the name of the great group preceded by one or more adjectives as in *Typic Cryorthods* – the typical Cryorthods. The system and its construction is shown below:

Order	Suborder	Great Group	Subgroup
S*po*dosols	*Or*thods	*Cry*orthods	*Typic* Cryorthods

A brief description of each of the orders is as follows:

Alfisols (Colour Plates IIIB and IVB): these soils have an argillic horizon and a high base saturation, they tend to occur on young landscapes in humid areas and equate with Luvisols.

Aridisols: these are the soils of the dry areas. They have a wide range of variability and include soils that have argillic, natric, salic, calcic, gypsic and cambic horizons. They include Xerosols, Yermosols and Solonchaks.

Entisols: these are the very young soils on recent sediments such as sand dunes and alluvium.

Histosols: these soils are composed predominantly of organic matter and include peats, mucks, bogs and moors. Also included are some thick aerobic organic materials.

Inceptisols (Colour Plates IB and IIA): these soils show a modest amount of development through the presence of a cambic horizon. They equate generally with Andosols, Cambisols and Gleysols.

Mollisols (Colour Plate IIA): most of these soils have a thick mollic horizon and include the Chernozems and Kastanozems.

Oxisols (Colour Plate IIB): these are the strongly weathered soils of the tropical and subtropical areas. They often have an oxic horizon and many have plinthite (laterite) and equate with the Ferralsols and Nitosols.

Spodosols (Colour Plate IVA): most of these soils have a spodic horizon, many have a brillant white albic horizon and some have a placic horizon. These include all the Podzols.

Ultisols (Colour Plate IA): these are the soils of the mid-to-low latitudes that have an argillic horizon and low base saturation. These soils are usually strongly weathered with high amounts of kaolinite and gibbsite. They equate with the Acrisols.

Vertisols: these are clayey soils that have deep cracks during the dry season. They are often dark in colour with high amounts of montmorillonite and a high base saturation.

FitzPatrick's system of soil designation and classification

The system gives designations or formulae for soils using a new horizon nomenclature. Two examples are:

$$Lt_2Fm_4Hf_3Mo_5Zo_5Sq_{50} - Asl$$
$$Lt_2Fm_3Hf_2Mo_7Zo_5Sq_{40}In_{20} - Asl$$

(Lt = litter; Fm = fermenton, 01 horizon; Hf = humifon, 02 horizon; Mo = modon, Ah horizon, ochric A horizon; Zo = zolon, E horizon, albic E horizon; Sq = sesquon, Bs horizon, spodic B horizon; In = ison, Cx horizon, fragipan; Asl = acid sandy loam parent material).

Formulae give the vertical sequences of horizons in the pedounit and their thickness in centimetres through subscript numbers. Such formulae give great flexibility in classification, allowing the creation of many different hierarchies to accommodate various users and disciplines. A pedologist may create a hierarchy based on the presence of the Sq, thus placing both soils in the same class. This would be the conventional grouping thereby creating a class of Podzols. An engineer might create a hierarchy based on the In to differentiate types of foundations, thus having two classes.

An added advantage is that combinations of two or more symbols can be used on maps, with the full formulae in the legend. Thus, unlike other systems, much information can be derived directly from soil maps.

The system is based on the idea of conceptual segments in space, created by using coordinates which are the individual soil properties. The segments are either reference (polar) or intergrade (continuum) segments. A reference segment, such as the sesquon, has a single unique and dominant property or a unique combination of properties formed by a single set of processes. An intergrade segment contains properties of two or more reference segments.

The above leads to the recognition of four fundamental types of horizon: reference or polar horizons, intergrade or continuum horizons, compound and composite horizons. Reference segments and reference horizons are given names based on some conspicuous property. The sesquon derives its name from the term sesquioxide with the *on* ending coming from horizon. Both reference segments and horizons are designated by the symbol Sq which is itself derived from the name sesquon. Intergrade segments and horizons are designated by combining the appropriate symbols in round brackets with the dominant placed first. The intergrade (AtSq) displays the properties of an alton (cambic B horizon) and a sesquon but is more like an alton.

Compound horizons contain the properties of two contrasting segments, one set of properties superimposed upon the other and caused by climatic change or progressive soil evolution. They are designated by placing the symbols in curly brackets thus {InFg}.

Composite horizons contain discrete volumes of two or more segments. In some Chernozems the horizon with crotovinas is designated (Ch/2Cz-) where Ch is the chernon and 2Cz- is the symbol for calcareous loess.

With this system any soil can be given a simple and clear designation with universal application.

8 *Geography of world soils*

The soils of the world are given using the FAO terminology, but in order to make the information more widely acceptable the terms of Soil Taxonomy (1975) and FitzPatrick (1980) follow in brackets where they differ from those of FAO.

The soils are presented mainly in the form of a transect; the first part is down through the USSR west of the Ural mountains and the second from the Sahara desert to Zaire. This treatment illustrates the broad geographical relationships between soils in two of the main continental areas of the world (Fig. 8.1). In addition a few important soils are mentioned that form as a result of maritime conditions or resulting from the local dominance of a particular soil forming factor.

In the north of the USSR there is the tundra with Gelic Gleysols (Cryaquepts, Cryosols) (Colour Plate IA and Fig. 8.10). These are characterised by a thin accumulation of organic matter at their surface followed by a dark greyish brown mixture of organic and mineral material below which is a wet mottled horizon about 50 cm thick. Then there is a sharp change into the permanently frozen subsoil or permafrost composed of thin bifurcating horizontal veins of ice surrounding lenticular peds of frozen soil. The upper horizons freeze every winter and thaw every summer, a cyclic process which causes expansion and contraction and the formation of surface patterns such as mud polygons, stone polygons (Figs. 3.4 and 3.5) and tundra polygons, which are often outlined by small depressions beneath which are vertical wedges of ice extending to a depth of 3 m or over (Figs. 8.2 and 8.3). The range of plant species in the tundra is restricted to a few mosses, liverworts, lichens, grasses and sedges but to the south, with higher temperatures, there are shrubs such as the arctic willow and finally spruce and larch forests.

Tundra soils are useless for agriculture because of the low temperatures, but they support large herds of reindeer and

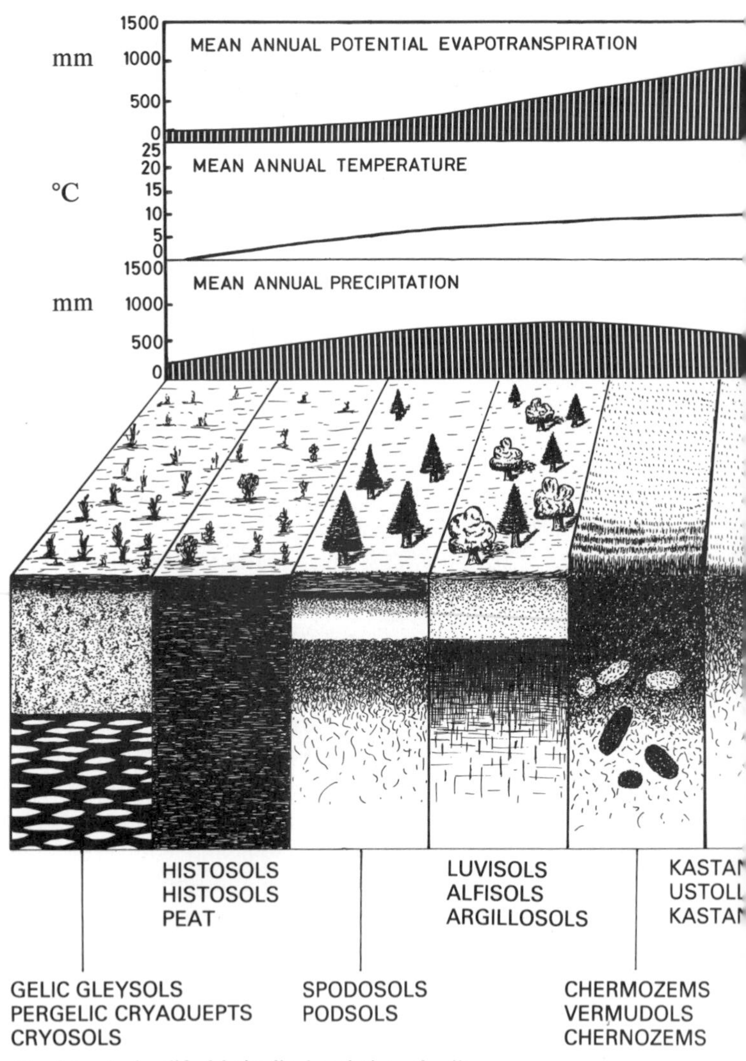

Fig. 8.1 A simplified latitudinal variation of soils

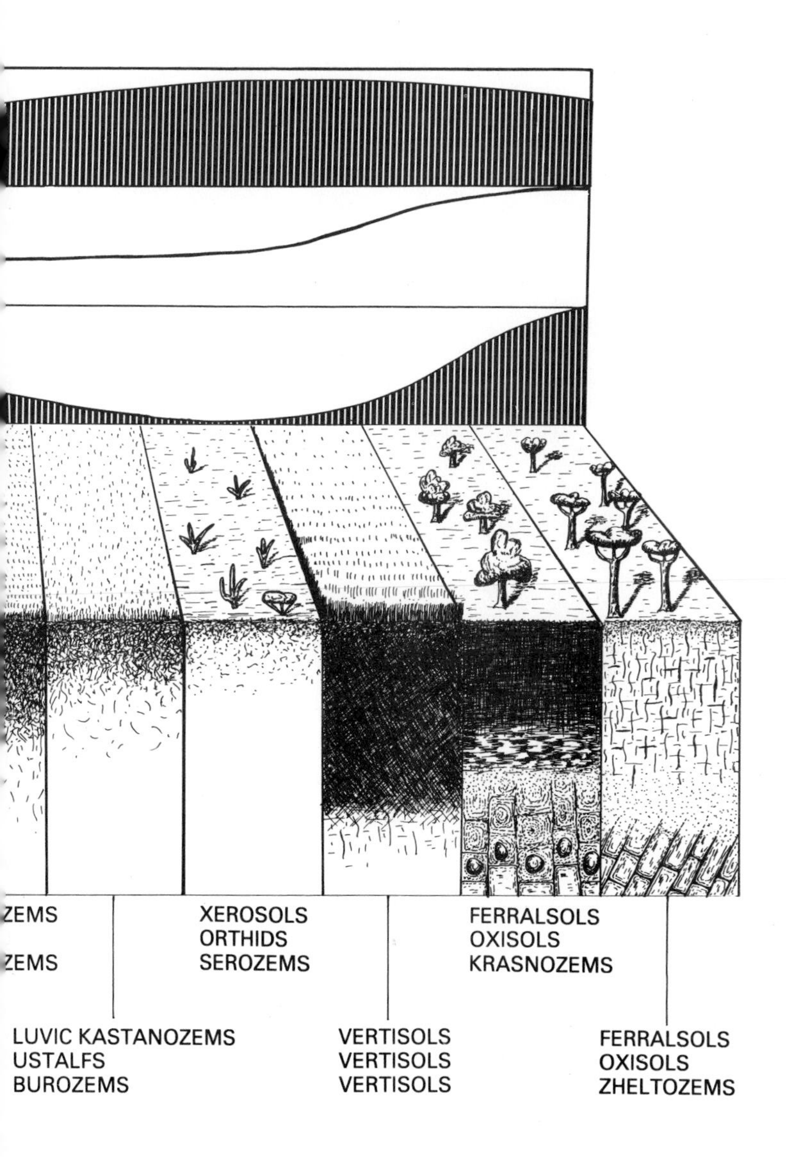
ZEMS
ZEMS
XEROSOLS
ORTHIDS
SEROZEMS
FERRALSOLS
OXISOLS
KRASNOZEMS
LUVIC KASTANOZEMS
USTALFS
BUROZEMS
VERTISOLS
VERTISOLS
VERTISOLS
FERRALSOLS
OXISOLS
ZHELTOZEMS

(A)

Fig. 8.2 (A) Oblique aerial view of tundra polygons in Northern Alaska. Each polygon is 15–20 m in diameter
(B) An ice wedge 4 m deep on Gary Island, northern Canada

caribou which either roam freely or are crudely managed.

South of the tundra there are extensive areas of Histosols (Histosols, Peat) (Figs. 8.1 and 8.11) with marshy vegetation dominated by *Juncus* spp., *Carex* spp. and mosses, particularly *Sphagnum* spp. Peat is an accumulation of organic matter under wet anaerobic conditions and is divided into basin peat and climatic peat. Basin peat forms in wet depressions and flat sites where the plant litter will not decompose rapidly but accumulates. Although the rate of decomposition of peat is slow, some changes do take place so that it ranges from very fibrous and woody to amorphous and plastic. Basin peat occurs elsewhere as in the Everglades of Florida and on the north coast of Borneo (Figs. 8.4 and 8.5).

Climatic peat forms in areas of high atmospheric humidity, low evapotranspiration and consequently wet soils. These conditions are found on the humid west coasts of Canada, Alaska, Britain and Scandinavia.

(B)

Basin peat receives much of its moisture by run-off which varies in composition and acidity. The fens of England are a good example of peat with high pH values due to the high content of calcium bicarbonate in the run-off water derived from the neighbouring chalk. On the other hand climatic peat is always acid because it derives its moisture from the atmosphere which contains very little basic cations and may contain acid pollutants.

Peat is the traditional fuel in many parts of the world but it is more valuable for horticulture and is cultivated extensively, but often needs to be drained and must be limed to raise its pH

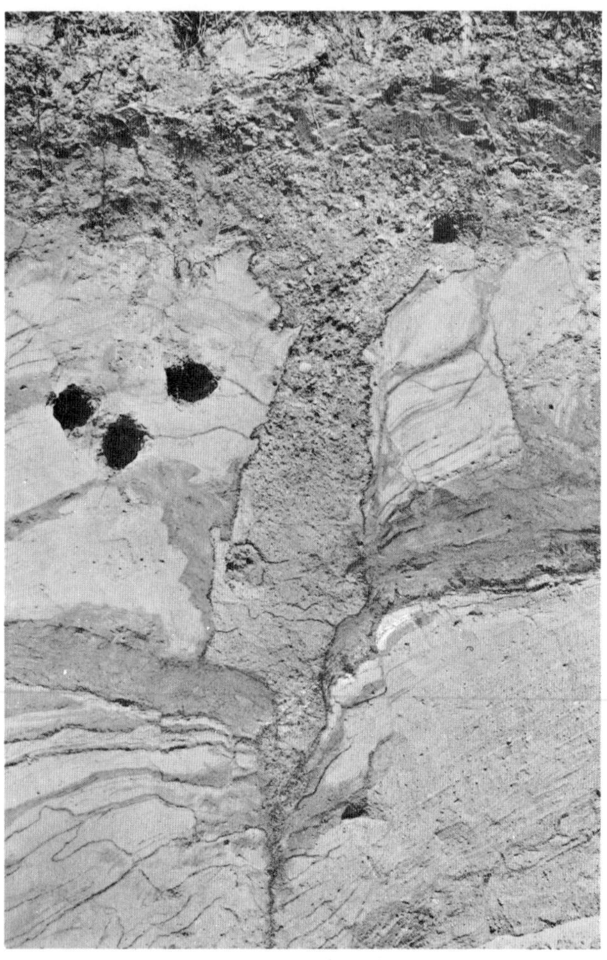

Fig. 8.3 Fossil ice wedge in Scotland. The vertical wedge which is 3 m deep is composed of sand and gravel but was solid ice when there was a tundra climate

(except for fen peats). In addition it often suffers from copper and zinc deficiency. The precise type of crop varies from place to place and include forests, permanent grassland and crop plants such as onions.

Great areas of coniferous forests follow the peat and here the soils are predominantly **Podzols** (Spodosols) (Colour Plate IVA

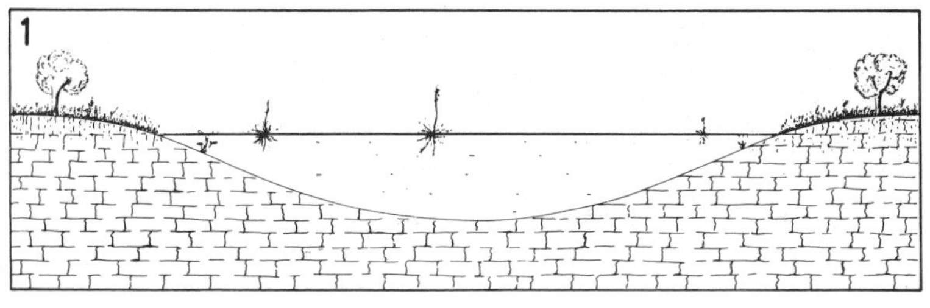

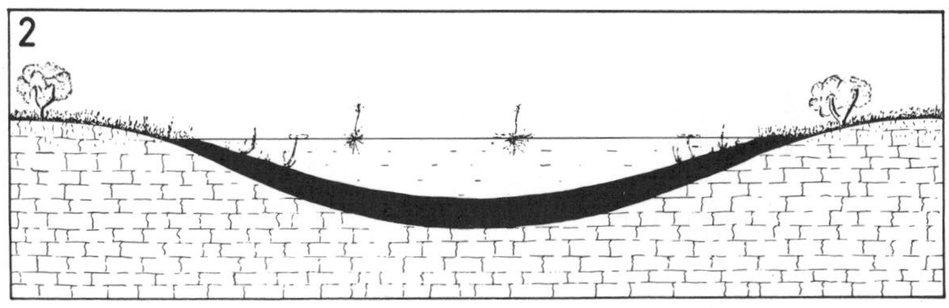

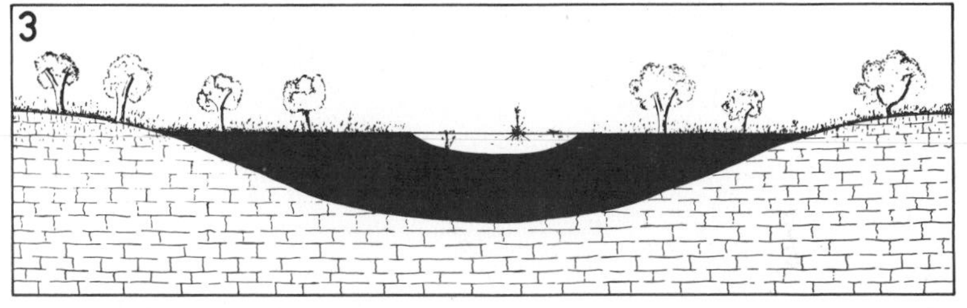

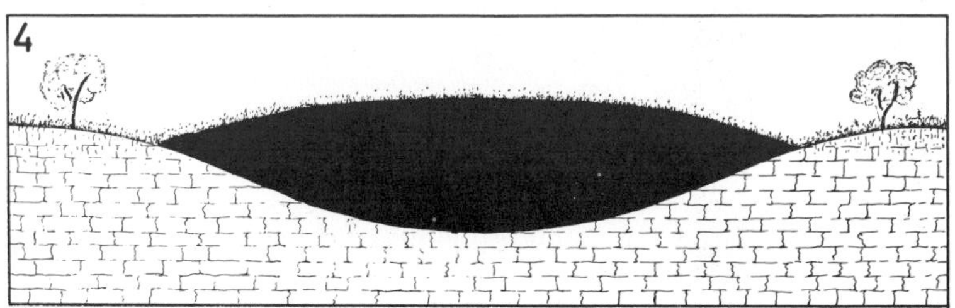

Fig. 8.4 Peat formation

1. A small pond or lake which could be situated between two moraines
2. At the bottom of the pond there is a thin accumulation of organic matter from the plants growing in the pond and in the surrounding soil
3. Considerable increase in the thickness of the organic matter and the spread of vegetation on to its surface
4. Continued thickening of the peat to develop the characteristic domed form of the final stage

Fig. 8.5 A peat deposit containing a layer of tree stumps that indicate a dry phase and forest cover during its accumulation

and Fig. 8.12) on the freely draining sites with the morphology and processes described in Chapter 1. There is a litter layer followed by progressively more decomposed organic matter then an organic mineral mixture, a bleached sandy horizon (albic E horizon, zolon), a horizon of sesquioxide and humus accumulation (spodic B horizon, sesquon) and finally the unaltered underlying material. The water percolating through the decomposing organic matter dissolves acids then it enters the mineral soil causing decomposition of the minerals and the release of ions such as calcium, potassium and iron. These are translocated downwards leaving behind the light coloured upper horizon and form the mottled horizon. This usually changes sharply into a brown or reddish brown horizon with some mottling, but generally there is less evidence of anaerobism. These soils have a fine texture or there may be an increase of clay with depth causing the reduced permeability and the water to accumulate. Thus Surface water Gleysols have their horizon of maximum reduction sandwiched between two aerobic horizons.

The principal property limiting the utilisation of Gleysols is their wetness, therefore they have to be drained, then they can be used for producing a wide range of crops. As stated above (Fig. 8.6) there is a topographic relationship between Podzols, Ground water Gleysols and Peat.

Placic Podzols (Placaquods, Placosols), distinguished by the presence of a thin continuous iron pan (placic horizon, placon) are very common in cool oceanic countries such as Britain and Scandinavia. They usually have a heath or moorland community dominated by various *Ericaceous* species, *Carices*, lichens and mosses. At their surface there is a thin litter followed by up to more than 20 cm of well humified material. Then there is an organic mineral mixture followed by an olive grey faintly mottled horizon (albic E horizon, candon), underlain by the thin hard continuous iron pan, which may rest on a variety of horizons but they are often strongly compacted.

The iron pan is impervious, so that water is held up in the upper part of the soil, causing the formation of the olive grey horizon and the accumulation of organic matter. In addition these soils are very acid and nutrient deficient which render them almost useless for any type of utilisation. Before crops can be grown these soils must be limed, fertilized and cultivated, but above all the iron pan must be broken to improve their drainage. This last operation requires expensive equipment which is not often available. At present considerable efforts are being made to establish coniferous forest on these soils, particularly in the British Isles.

The very fertile **Cambisols** (Ochrepts, Altosols) (Colour Plate IB and Fig. 1.2) occur beneath some deciduous woodland of the cool temperate areas. At the surface there is a loose leafy litter resting on a brown granular horizon containing numerous earthworms. Below is a brown, friable, middle cambic B horizon, (alton) which grades into the unaltered material that is often basic or calcareous, imparting a high base saturation and slightly acid to mildly alkaline pH values. These properties encourage earthworm and bacterial growth thus causing a rapid breakdown and incorporation of organic matter into the soil. Thus these soils have a high natural fertility and the progress of early humans through many parts of central Europe was such that they first cleared the forests from Cambisols and cultivated them and only when the pressure on the land became high did they turn their attention to the less fertile soils such as Podzols. The forests were oak and beech which were very valuable as building materials so that Cambisols provided the timber for houses and a suitable seed bed for crops. However, Cambisols need lime and fertilizers to support continuous cultivation.

There is a continuous gradational sequence from Cambisols to Podzols in which the soils become progressively more acid and leached. In some cases Cambisols can be changed into Podzols by planting conifers on them.

In many cool temperate areas of the world with deciduous forests or mixed deciduous and coniferous forests there are **Luvisols** (Alfisols, Argillosols) (Colour Plate IIIB). These soils have a middle horizon containing more clay than the horizons above or below, as a result of translocation of clay from the upper horizons (argillic B horizon, argillon). Thus the profile of Luvisols has a thin loose leafy litter resting on a greyish brown mixture of organic and mineral material. This changes into a grey sandy horizon followed by a sharp change into the brown blocky or prismatic horizon which has the high content of clay, then there is a gradation to the unaltered material.

Luvisols are weakly acid at the surface with medium base saturation which imparts a moderate natural fertility but they need lime and fertilizers as a normal part of arable cultivation. At present they are very productive, supporting a wide range of crops and various types of animal husbandry, particularly dairying, because they occur in areas with a high level of technology.

A very characteristic and distinctive feature of soils is their gradual change from one to another. Fig. 8.7 attempts to illustrate some of the many changes that occur in the landscapes of temperate areas. The Cambisol in the middle is shown as having three principal horizons, an upper organic mineral mixture (Ah,

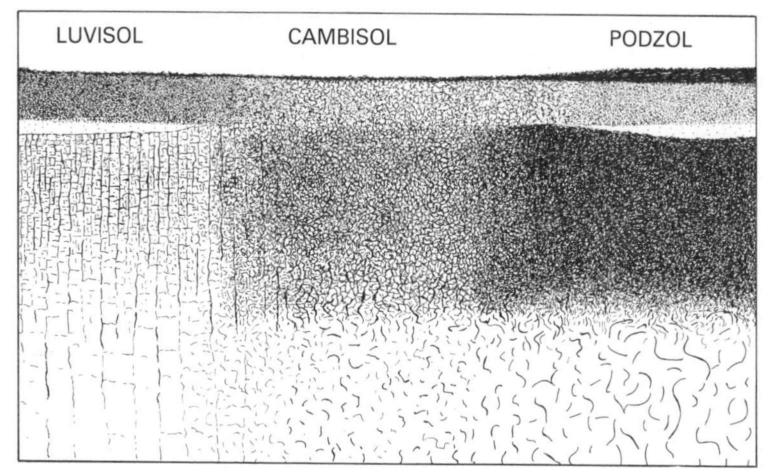

Fig. 8.7 Lateral gradation of soils and soil horizons

umbric A horizon, mullon) a middle brown horizon (Bw, cambic B horizon, alton) and the relatively unaltered underlying material. To the right these horizons gradually change through intergrade horizons into a Podzol with an upper bleached horizon (E, albic E horizon, zolon) a middle horizon (Bs, spodic B horizon, sesquon) containing material removed from the horizons above and the underlying material. To the left the Cambisol changes into a Luvisol. The upper organic mineral mixture changes into an ochric A horizon while the middle horizon intergrades into one containing clay washed down from the horizon above (Bt, argillic B horizon, argillon).

South of the deciduous forests precipitation falls to less than 400 mm and there is a zone of tall grass growing on **Chernozems** (Mollisols) (Colour Plate IIA and Fig. 8.13) which is the Russian for black soils. These soils have a root mat at the surface resting on a thick black horizon up to 2 m in thickness (mollic A horizon, chernon). Within the lower part of this horizon and below there are thin thread-like deposits (pseudomycelium) of calcium carbonate and also there may be concretions of calcium carbonate. The black horizon grades into the yellowish brown underlying material which is usually loess.

A conspicuous feature of European Chernozems is the presence of burrows (crotovinas) caused principally by the blind mole rat. These crotovinas extend into the underlying material and are seen as dark infillings in the lighter coloured underlying material or as light areas in the dark top soil. In addition these soils have a vigorous earthworm population. Because precipitation is <400 mm, only enough water passes through the soil to remove the most soluble salts and to translocate calcium carbonate, hence the pseudomycelium and concretions. These soils have pH values about neutrality, very high base saturation, and a very high inherent fertility. They form the main areas of grain production, particularly wheat and more recently maize. Initially it was thought that they had an inexhaustible fertility but after decades of cultivation, production has gradually declined so that it is now customary to add fertilizers.

In the USA and elsewhere there are **Phaeozems** (Udols, Brunizems) which are similar to Chernozems but they have a middle horizon with a clay maximum and are slightly less fertile.

South of the Chernozems the precipitation gradually decreases from 400 mm to <100 mm and the vegetation changes from tall grass, to short grass, to bunchy grass, and then into species that will withstand long dry periods. There are parallel changes in the soils and gradually the thick black horizon becomes lighter in colour and shallower and the horizon of calcium carbonate comes nearer to the surface but generally three distinct soil types can be recognised, viz., Kastanozems, Xerosols and Yermosols. **Kastanozems** (Ustols) have a short grass vegetation and dark brown horizon up to about 30 cm thick, below which is the horizon of calcium carbonate accumulation followed by the unaltered material. **Xerosols** (Aridisols, Serozems) have a bunchy grass vegetation and a thinner brown upper horizon then the lower calcium carbonate horizon. **Yermosols** (Aridisols) have typical desert type vegetation and a thin brownish grey upper horizon followed by that of carbonate accumulation.

These three soils have a very high inherent fertility but productivity is usually restricted by the lack of moisture. This can be overcome by irrigation or moisture conservation but often in these areas there are only a few rivers or other sources of irrigation water. An exception is the Colorado river which runs through the desert in the south western part of the USA.

When irrigation is impossible dry-farming is carried out particularly in areas of Kastanozems. In areas of Xerosols and

Yermosols water conservation by dry-farming is inadequate to support crop growth, irrigation is essential.

Within the semi-arid areas where evapotranspiration greatly exceeds precipitation, certain cations and anions accumulate in the soils to cause high salinity or alkalinity. The ions include sodium, potassium, magnesium, calcium, chloride, sulphate, carbonate and bicarbonate. Since the upper limit for salt tolerance by most plants is about 0.5 per cent, soils containing higher amounts are regarded as being saline and generally referred to as **Solonchaks** (Salorthids) and are usually easily recognised by salt efflorescences on their surfaces (Fig. 4.18). The soil profile often resembles a Gleysol soil by having an upper grey organic mineral mixture which rests on a mottled horizon followed by a grey or olive completely reduced horizon.

Many Solonchaks are potentially useful for agriculture if the excess ions can be dissolved and removed. This requires large volumes of water but sometimes a suitable supply is not available because of the general lack of adequate amounts of salt-free water in these dry areas.

In some soils of dry areas a water-table with a large amount of dissolved salts may be present at depth. If such soils are over irrigated, as is often the case, the irrigation water reaches down to the water table and may cause it to rise to the surface by capillarity, thereby inducing salinity and severely reducing crop production. This has happened in many areas, a good example being the Sind valley in Pakistan where thousands of hectares are lost to agriculture every year (Fig. 8.8).

Solonetz (Natrustalfs) also occur in semi-arid areas and have at their surface a thin litter followed by a thin very dark mixture of organic and mineral material and then a dark grey somewhat sandy horizon. Below is the very distinctive middle horizon (natric B horizon, solon) with its marked clay increase, characteristic prismatic or columnar structure and a pH value that is often over 8.5 due to high exchangeable sodium and magnesium but generally a low content of salt (Fig. 4.10 and 4.15). The high pH is extremely harmful to plant growth and must be lowered before cultivation can be carried out. This is usually achieved by adding calcium sulphate, the calcium enters the exchange complex, replacing sodium which is removed by leaching following natural rainfall or irrigation. The calcium has other

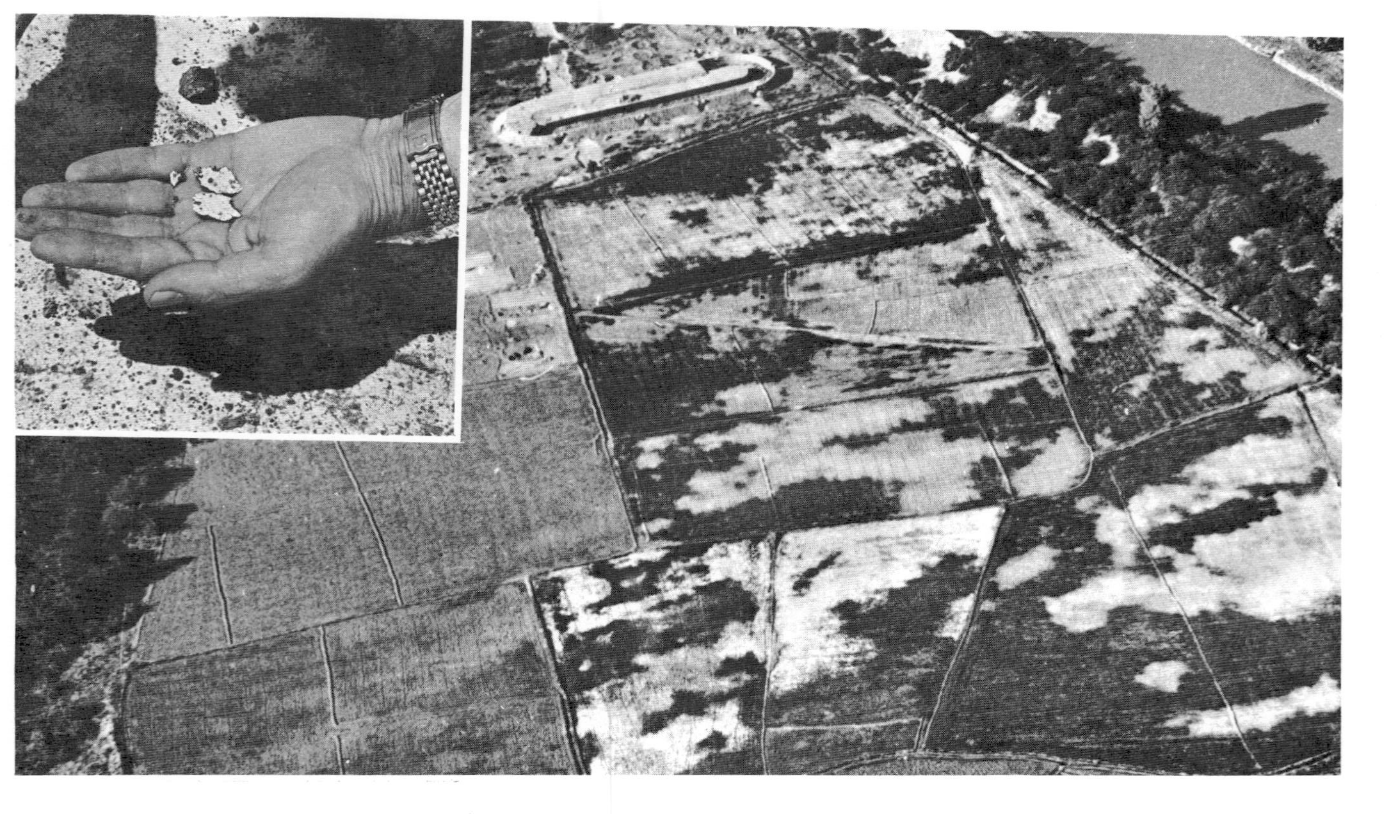

Fig. 8.8 Salt spreading in farmland in Pakistan. The inset shows the salt crust that forms on the surface

beneficial effects for it improves the structure and is an essential element for plant growth and microbiological activity.

Solodic Planosols (Natraqualfs, Solods) can be regarded as leached Solonetz in which the upper horizons are strongly bleached becoming pale grey or white. The middle horizon has a clay maximum, high exchangeable sodium and/or magnesium, but it is acid. When adequate amelioration has been carried out these soils can be used for growing a wide variety of crops.

Deserts occupy about one fifth of the earth's surface and in the Sahara vast areas are rocky, stony or in the form of sand dunes. The areas of loose material are not really soil but they have a high potential for crop growth if irrigated. However, the intense solar radiation causes plants grown under these conditions to transpire very rapidly during the day, leading to physiological drought and often permanent damage. Therefore, desert areas in many cases offer a low potential for crop production.

Within semi-arid regions of the tropics and subtropics there are extensive flat areas with deep, dark coloured, clay soils dominated by montmorillonite generally known as **Vertisols** (Fig. 8.14). The surface horizon is usually granular but can be massive and is followed by a dense horizon with prismatic or angular blocky structure that grades into similar material with a marked wedge structure. This is caused by pressures that develop as a result of expansion and contraction of the montmorillonite clay in response to wetting and drying. Many of the peds have slickensides i.e. shiny surfaces which form as one ped slips over another during expansion.

Vertisols usually have pH values about neutrality, a high base saturation and high CEC, therefore they have a high potential for agriculture but often suffer from drought; but under irrigation their productivity is very high. Perhaps one of the most famous areas is the Gezira of the Sudan where there is a high output of excellent quality cotton which forms the backbone of the country's economy.

Within the humid tropics are highly weathered soils that are often bright red in colour and known as **Ferralsols** (Oxisols, Krasnozems) (Colour Plate IIB and Fig. 8.15). These soils have a thin litter at the surface followed by a greyish red mixture of organic and mineral material which is not more than 5–10 cm thick. This grades quickly into a bright red horizon (oxic B horizon, krasnon) which may be several metres thick followed by a change into the underlying rock. This red horizon is usually composed largely of kaolinite with some iron oxides but there may be a very high content of gibbsite. It is then regarded commercially as bauxite and mined for the production of aluminium. The change to the underlying rock can be gradual through progressively less weathered rocks or it may be sharp. Often there is a characteristic red and cream mottled horizon beneath the red top soil or there may be a thick white horizon known as the pallid zone. In many of these soils there is a vesicular slag-like horizon generally known as laterite which may be soft within the soil but hardens upon exposure.

Ferralsols are formed by the progressive weathering of the rock until all of the primary minerals such as feldspars, amphiboles and pyroxenes have been completely decomposed and transformed into secondary substances or lost from the system, hence the fact that the middle horizon has a high content of kaolinite or oxides (Fig. 3.15). Because weathering has been so complete these soils are very deficient in essential plant elements but they often carry very luxuriant high forest because the elements are constantly being recycled and there is a small reservoir in the top few centimetres due to their release by the decomposing litter. This caused many earlier workers in the tropics to be misled for they assumed that the high forests were indicative of very rich soils. In the majority of cases when the high forest is cut and agriculture attempted there is failure within two or three years, during which the essential plant elements in the surface are exhausted. This is known to the native people who practised a system of shifting cultivation. They cleared an area and cultivate it for two or three years and then move on to a fresh site allowing the depleted site to develop a secondary forest and for the reservoir of fertility to be rebuilt. When fertilizers are available these soils respond well to cultivation and grow a variety of crops such as cocoa, coffee, sugar cane and oil palm.

Because of the wide climatic fluctuations during the Pleistocene period many Ferralsols that developed under humid tropical conditions are now in another environment. This is particularly marked in Australia where red soils and laterite occur commonly in the desert and elsewhere.

Although Ferralsols are common soils within the humid tropics there are other soils such as Ferric Cambisols that are shallow and not as strongly weathered and contain primary minerals.

Some of the most dramatic changes brought about by humans are seen in some tropical and subtropical countries as a result of growing rice. Many brightly coloured soils now have drab and mottled colours due to the long periods of waterlogging required by the rice plants.

In a number of tropical and subtropical areas with a marked dry season such as Western Australia, there are distinctive yellow and yellowish red soils containing an abundance of subspherical concretions about 1 cm in diameter. These soils are about 3 m thick and are usually loose at the surface but may be cemented at a depth of 1–2 m. Below is the zone in which the concretions are forming and overlying the weathered rock (pallid zone). Although these soils are very common in some places they have not been named, nor have the processes of formation been studied. Because the dominant property seems to be the presence of subspherical concretions they might be called **Spherosols** (Fig. 8.9).

In West Africa, Australia and apparently also in South America it is very common for a very coarse textured horizon to occur just beneath the surface organic-mineral mixture and overlying a cream and red mottled horizon on weathered rock. These soils have been formed by differential erosion during which the clay has been removed leaving the sand, gravel and concretions behind. Current research indicates that they are probably among the most common soils in many tropical and subtropical areas and might be called **Cumulosols**.

Arenosols (Psamments) are composed predominantly of coarse sand dominated by quartz and occur in many tropical and subtropical areas with a marked dry season. They may be several metres thick with uniform colour and range from bright red to yellow and even pale brown or grey depending upon climatic conditions. They can develop from sandy parent material or they may form by the differential removal of fine material leaving a concentration of sand behind. These soils are inherently infertile. However, they can produce a number of crops if fertilizers are applied but moisture is often the principal limiting factor.

Andosols (Andepts) are formed mainly on volcanic ash. At the surface is a loose little followed by a dark coloured mixture of

Fig. 8.9 Concretions in a Spherosol

organic and mineral material which changes fairly sharply into a brown middle horizon (cambic B horizon, andon) that grades into the parent material. The unique property of these soils is their high content of allophane which gives them a very low bulk density and fluffiness particularly in the middle horizon. Probably the greatest extent of these soils is in Japan and New Zealand.

Two distinctive soils that develop from limestone are **Rendzinas** (Rendolls) and **Chromic Cambisols** (Tropepts, Rossosols) and generally known as **Terra Rossas**. Limestone is an exceptional rock, being composed largely of soluble calcium carbonate which is dissolved during weathering so that only the very small amount of impurities in the rock provide the residue to form soils, which as a consequence are never very deep.

Rendzinas have an upper black or very dark brown granular mollic A horizon (mullon) speckled with white fragments of limestone. This usually does not exceed 50 cm and may change abruptly into the white limestone or there may be a narrow transition composed of the black upper horizon and larger fragments of limestone. These soils are usually extremely fertile with pH values about neutrality, but may suffer from drought during dry years because of their small volume for water storage (see Fig. 6.1). For this reason many Rendzinas are not cultivated but have a natural or semi-natural vegetation such as the beechwoods of the Chiltern Hills in Southern England and numerous parts of the Mediterranean.

Terra rossas are characteristically brilliant red in colour and commonly are more than one metre in thickness. They are very old soils and represent the end stage of soil formation on limestone during which all of the carbonate has been removed, just leaving behind the accumulation of impurities. They occur in many humid parts of the world but they are principally associated with the Mediterranean where they are used extensively for growing grapes. This area, having been cultivated since prehistoric times, means that these soils are probably among the world's most eroded soils – a feature that was noted by Plato.

Bordering most of the world's major rivers are soils developed in deposits of alluvium and known as **Fluvisols** (Fluvents). In a number of cases these soils are extremely fertile and sustain a very high level of agriculture. It was upon the alluvial soils of the Nile, Tigris and Euphrates rivers that the first major western civilisations developed. This is not a mere coincidence for without fertile soils and an abundant supply of food, healthy and thriving communities cannot develop.

Although the transects through other continental land masses such as North America sometimes give similar sequences there are often major differences. For example in Australia deeply weathered rocks can be found throughout most of the country and within the most arid parts of the interior. This weathering probably took place during a much wetter climatic phase of the Tertiary period and now with a change to aridity the soils have many characteristics of those found in a more humid environment. Thus they display a unique polygenesis.

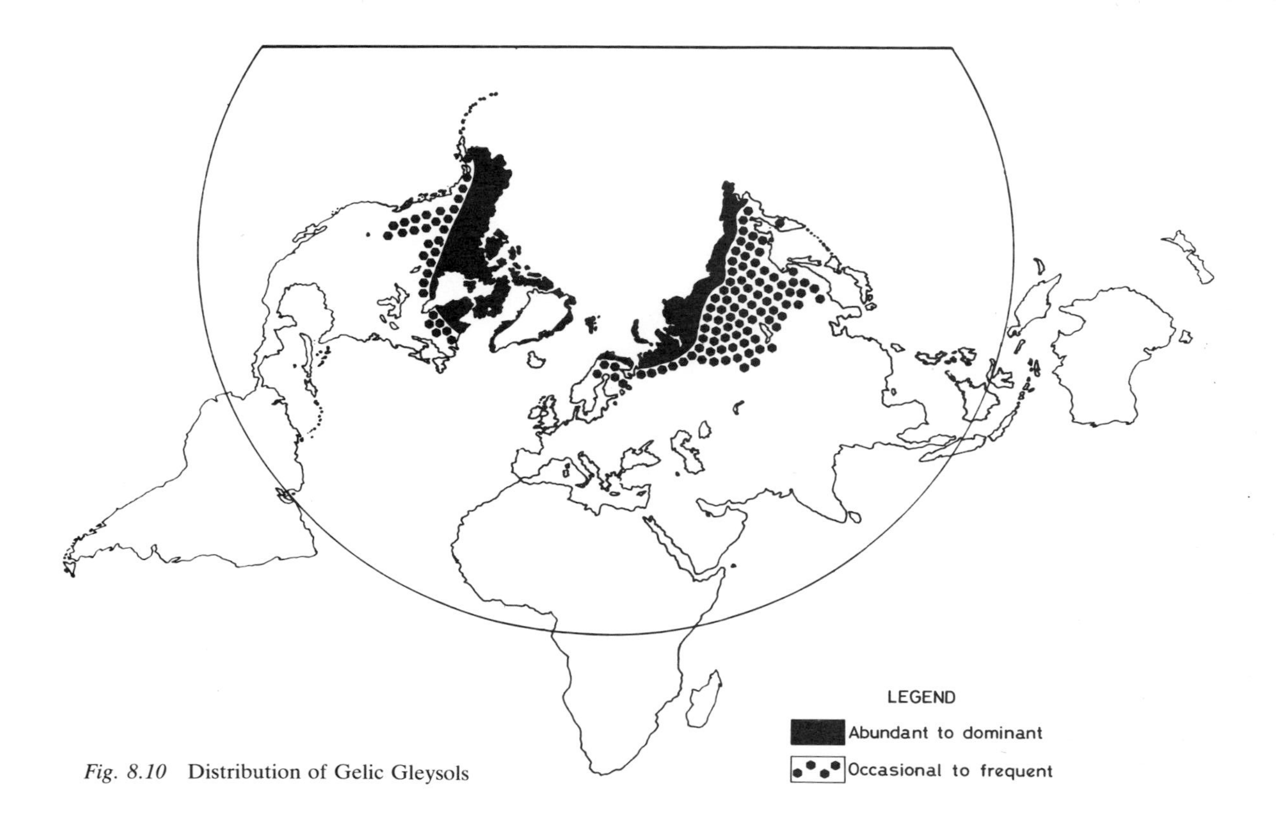

Fig. 8.10 Distribution of Gelic Gleysols

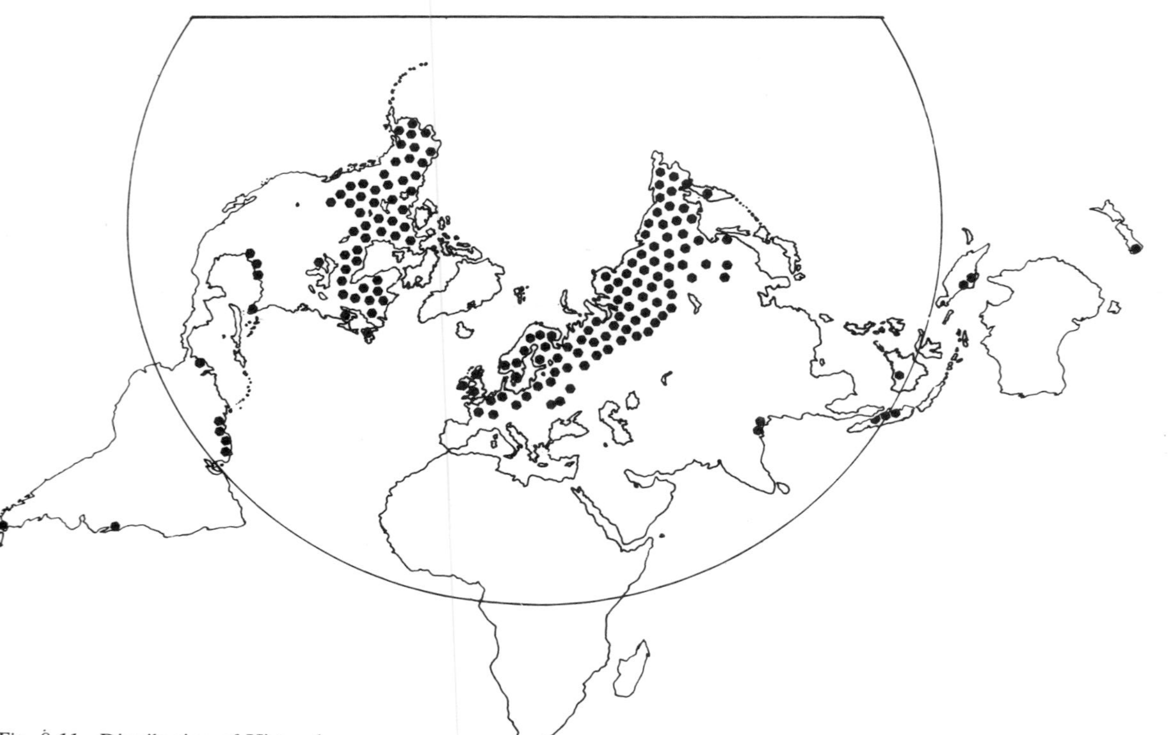

Fig. 8.11 Distribution of Histosols

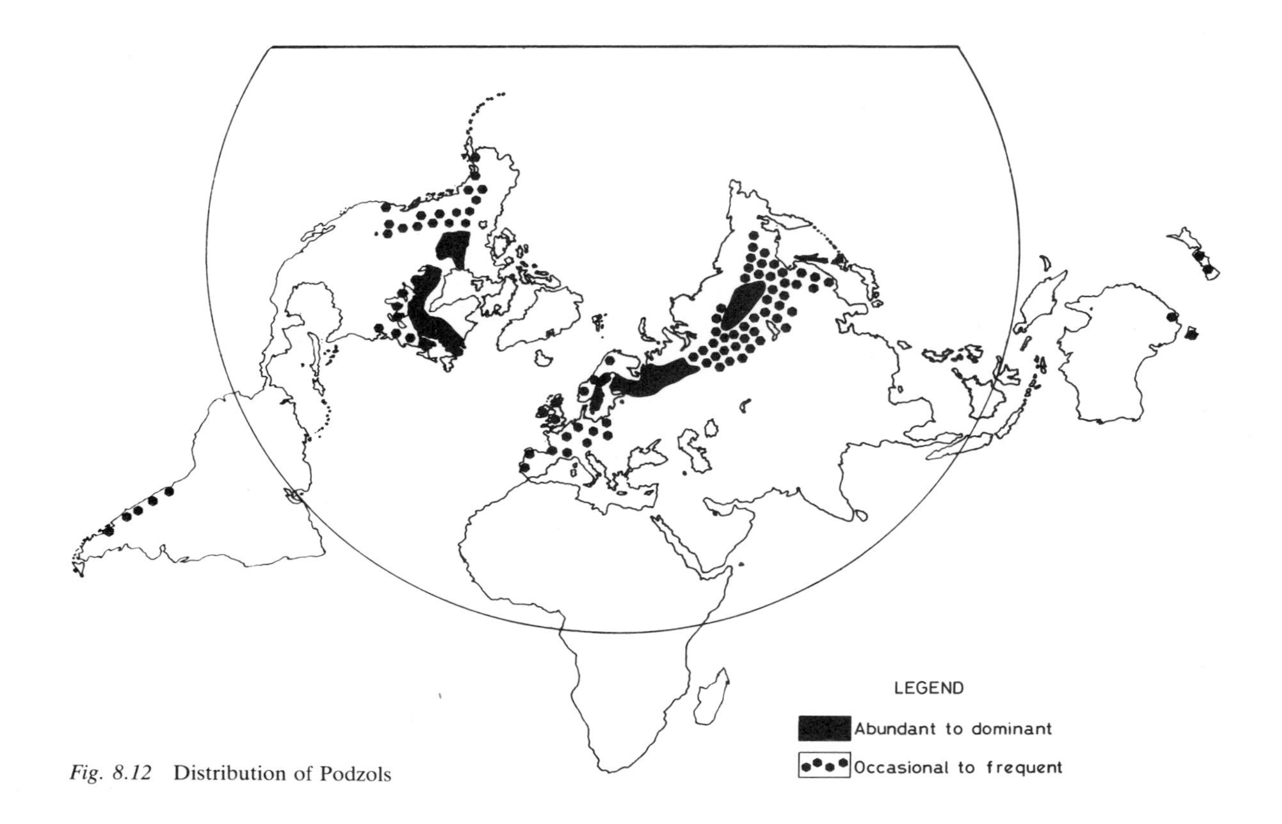

Fig. 8.12 Distribution of Podzols

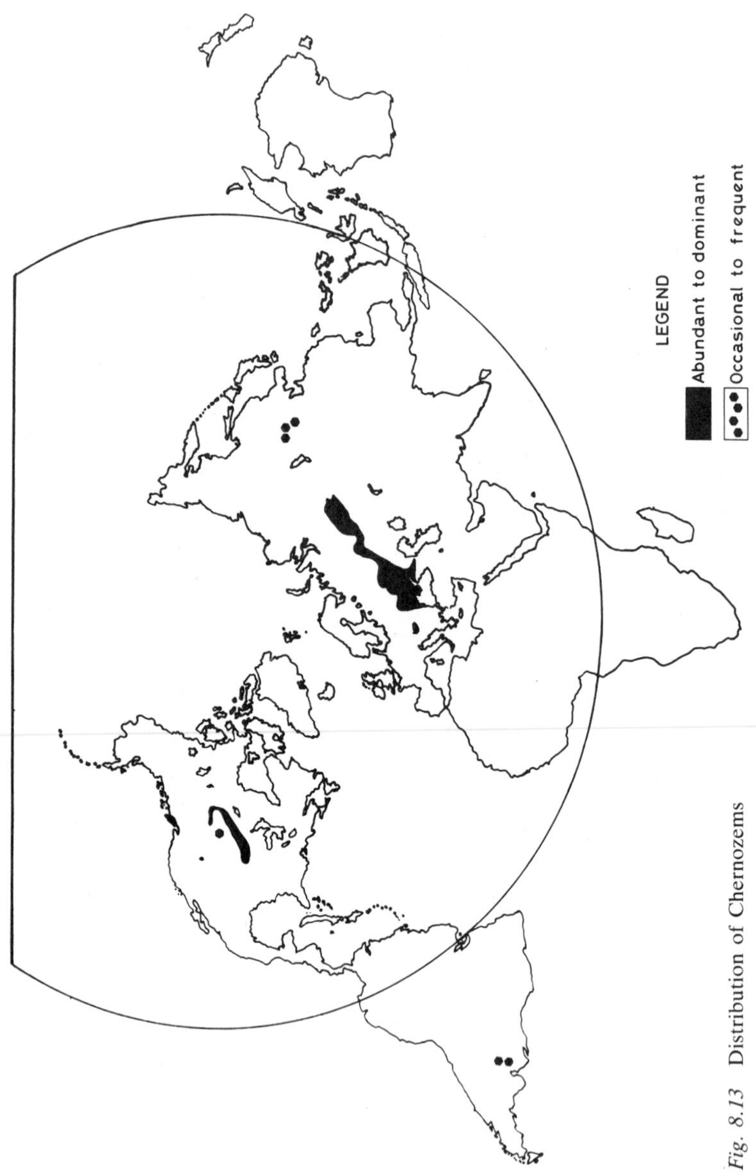

Fig. 8.13 Distribution of Chernozems

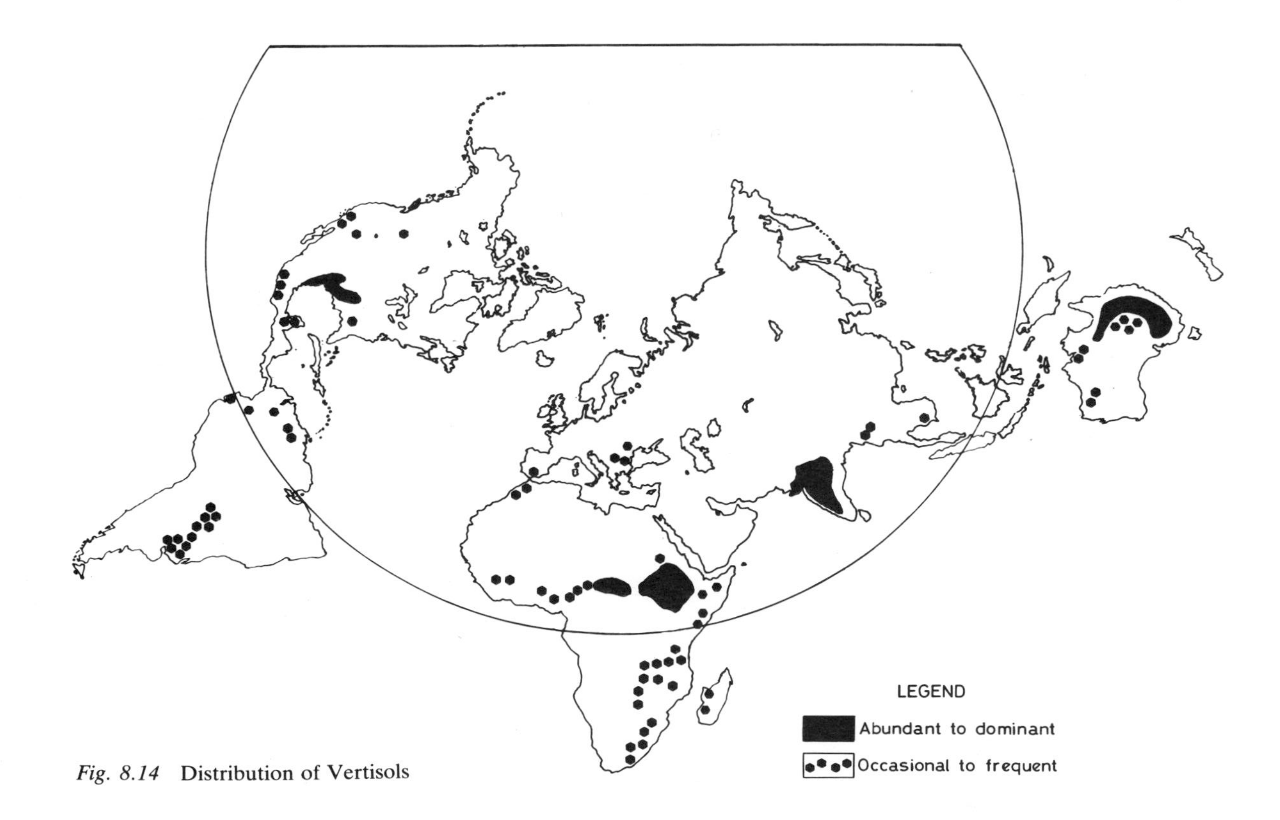

Fig. 8.14 Distribution of Vertisols

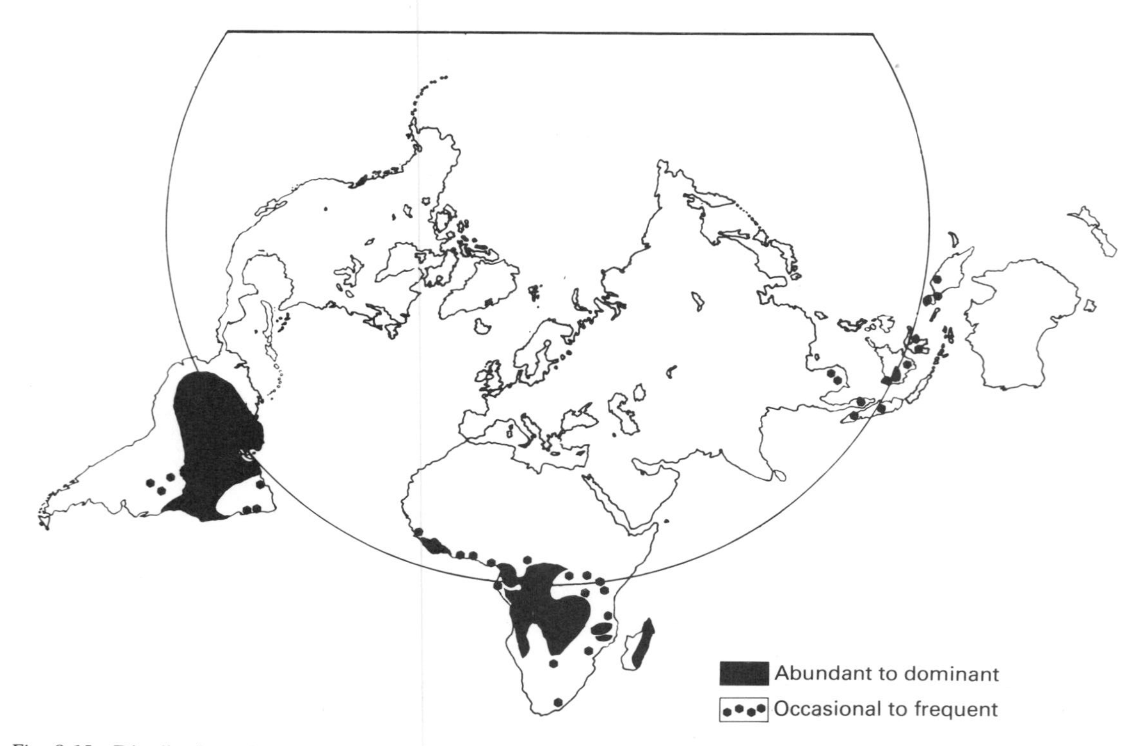

Fig. 8.15 Distribution of Ferralsols

9 *Soil maps and mapping*

One of the major occupations of soil scientists is surveying and mapping soils, and the production of soil maps. Surveying soils is a most exacting scientific task because soils do not normally have sharp boundaries but gradually grade from one into another. Further, this gradation is seldom clearly expressed on the surface of the ground and has to be determined by making numerous examinations which take the form of small inspection pits or auger holes.

At the outset of a soil survey it is essential to establish the purpose of the map: it may be a special or general purpose survey. In a general purpose survey the soils are mapped according to their morphology. If it is a special purpose survey such as for irrigation or degree of salinity then a restricted and specific number of properties are mapped.

Soil surveys usually take months or years to complete thereby involving considerable expense, therefore careful planning is necessary. This includes getting suitable staff, equipment, accommodation, transport, maps and laboratory facilities.

The soil survey procedure includes:

1. Collecting and studying all relevant data (geology, vegetation, maps. etc.)
2. Studying aerial photographs and delimiting preliminary boundaries
3. General reconnaissance if possible
4. Surveying and mapping, including descriptions of present and previous land use
5. Describing and sampling representative soils
6. Laboratory analyses of samples
7. Preparing map(s), and report(s)

The initial part of the survey usually starts with an examination of the aerial photographs of the area and from which preliminary boundaries may be drawn then to be checked by field examinations.

It is also important at this stage to determine which properties are to be mapped. This leads to the choice of classes to be mapped and the map legend and where necessary to establish any relationships between mapping units and land use planning.

The soil units that are mapped vary with the purpose of the map and the nature of the soil pattern. Generally the soil surveyor delimits *soil series* which are areas of relative uniformity but because soils are so variable soil series are seldom absolutely uniform and may contain up to 15 per cent of other soils.

There are various methods for conducting soil surveys. Most are "free surveys" based on the experience of the soil surveyor who criss-crosses the area making inspections at points in the landscape where it is considered necessary to determine the nature of the soils and to establish boundaries. At each point of inspection the properties of the soil are recorded on a map and when necessary in a notebook. Alternatively, one carries out inspections at predetermined locations which are based on some type of statistical program. This technique is more objective but is often time consuming; however it has to be used in areas for which there are no maps or aerial photographs.

The type of map on which the data are recorded can vary very much from place to place. In countries such as Britain, ordnance survey maps are used, but for many places such maps have not been produced. Then it is customary to use aerial photographs and to record the data either directly on to the photograph or on to a transparent overlay. Aerial photographs have a number of advantages, particularly with regard to finding one's location but aerial photographs have a varying scale across the photographs; therefore they cannot be used as complete substitutes for maps. The scale of the map varies from the FAO/UNESCO world soil map at 1:5,000,000 to as large as 1:10,00 for detailed maps in which individual fields are shown.

From the data recorded on the map or photograph and in the notebook the surveyor draws lines on a map or photograph to enclose areas of relative uniformity and so to produce a soil map. It may be necessary to group soils that are in some way related. In some cases all the soils with the same type of parent material are grouped. In areas with a repeating topographic pattern the grouping is based on topography. This latter type of grouping which is sometimes referred to as *land system* is very popular because a considerable amount of mapping can be based on aerial photography with a limited number of ground checks. With this system it is assumed that there is a correlation between topography, geology, vegetation and soils.

The field map is then submitted to the drawing office for the preparation of the final coloured map and its accompanying legend.

Most soil survey data is now computer stored. This includes the field data as well as the soil descriptions and analytical data. This allows rapid retrieval of information and the production of special purpose maps to suit user requirements. Published maps have a colour scheme based on the major soil groups. For example in Britain Podzols are in various shades of red, Gleysols in blue, Peat in purple, etc. Accompanying the soil map is a report which gives further details about the soils particularly some of their chemical and physical characteristics. Fig. 9.1 is an extract from a much larger soil map. It shows the distribution of the soils and the spatial relationships between the various soils. The distribution pattern is determined largely by slope and elevation. The Podzols occur >500 m on a plateau surface. On the slopes leading down from the plateau are Cambisols while flush gleys occur mainly at the break of slope between the Podzols and Cambisols. In the flat low-lying areas there are surface water gleys. In a number of reports there are diagrams which not only show the relationship between the soils but also relationships with slope and elevation. These are known as block diagrams such as shown in Figs. 1.3, 2.19, 2.31. The aerial photographs given in Figs. 9.2 and 9.3 are examples of areas where the different types of vegetation are strongly correlated with different soils. From the original soil map it is possible to derive a number of other maps and it is becoming customary to prepare a land capability map which is published together with the soil map and report. Although areas of soils can have a variety of uses, the land capability map is usually produced for agricultural purposes but they can be produced for special purposes such as housing or industrial development. Bibby & Mackney (1969) have produced a system of land use classification for the British Isles.

Soils are the world's major natural resource and a soil map is the spatial representation of these resources. Therefore soil maps are fundamental and should be the starting point when planning

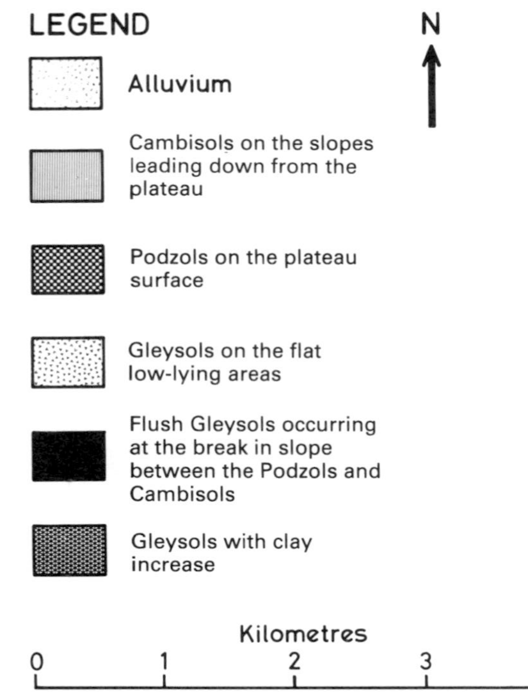

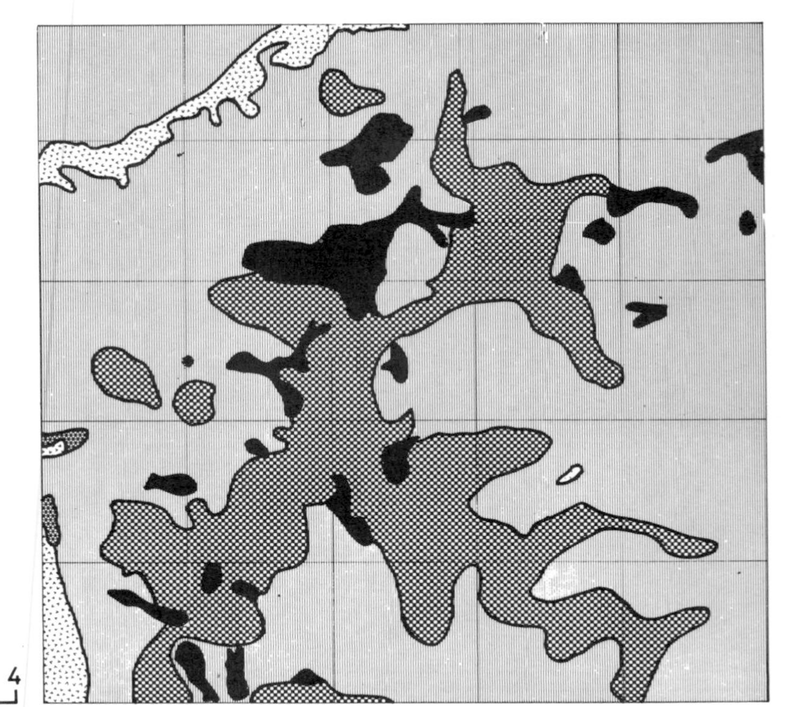

Fig. 9.1 Soil map of the area west of Church Stretton in Central England

Fig. 9.2 Salt marsh and reclaimed land in Norfolk, England. Salt marshes are generally areas of silt accumulation which can be reclaimed by constructing a bank and a drain on the landward side of the bank, the material dug out of the drain being used to make the bank. The enclosed creek system is then rationalised, a few larger creeks are maintained as major drainage channels and the small ones filled in.
The distinctive pattern on the marsh is due to differences in the distribution of the natural vegetation, creek pattern and reclamation.

A. Dominated by the light coloured cordgrass with sea poa as the dark areas
B. Dominantly sea poa
C. Clearly defined well drained areas adjacent to the creeks and dominated by Sea-couch
D. The bank
E. The drain
F. The pattern of infilled creeks where crop-growth is poor

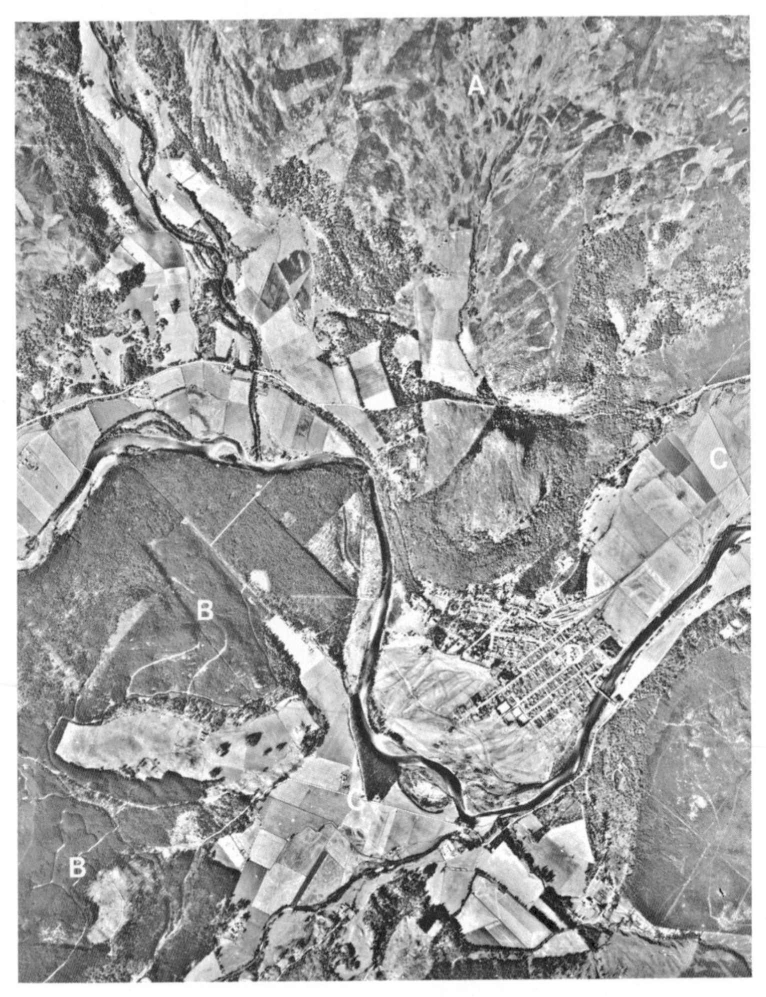

Fig. 9.3 Land use pattern around Ballater, Scotland. This is a good example of the wide range and strongly contrasting systems of land use in Highland Britain

A. The highest ground above 300 m has a patchy pattern produced by burning the heather which regenerates to produce young shoots for grouse that are shot for pleasure
B. At a lower level are forests dominated by Scots pine
C. The flat cultivated alluvial terraces form the lowest part of the landscape
D. The town of Ballater is situated on alluvium and partly surrounded by a loop in the river Dee

every type of land use whether it be for agriculture, forestry, road building or house construction. It is essential that the land surface should be utilised to its optimum, for example, good agricultural land should be kept for agriculture and not used for house building. A particularly striking example of the misuse of land occurs near London where Heathrow Airport was constructed on some of the best soils in Britain. Although soils may be a major factor in determining land use, economic and sociological factors may in the end be deciding factors.

In the last few years there has been a very rapid development in remote sensing techniques, a few of which are now used in a routine manner in soil and resource surveys. These include:

1. Normal aerial photography
2. Infrared photography
3. False colour photography
4. Multispectral scanning

Aerial photography mentioned above is the main type that is used. The photographs may be black-and-white or colour. The former are mainly used since colour although very pleasing to the eye does not seem to give much more information than black-and-white photographs.

Infrared photography will record the same image as normal panchromatic film plus the near infrared radiation, thus increasing the tonal contrast since water strongly absorbs infrared radiation while it is strongly reflected by vegetation. Therefore this film is of value where soils with varying moisture contents are being investigated.

False-colour photography produces colour images but without their normal or familiar colours. This is necessary to record the infrared band visibly and in colour. This is achieved by using an infrared colour film which shifts all the colour bands to a shorter wavelength. Blue is filtered out, green is printed as blue, red as green and near infrared as red. Thus vegetation appears red, bare soil appears blue-green, whilst wet areas are dark.

Multispectral scanning records the reflection of sunlight from the Earth's surface in a number of spectral bands, two within the visible and two near-infrared. In this technique the reflection from the ground is scanned by rotating mirrors and reflected onto a detector and recorded as a continuous signal on tape. There are three ways of using the data. The signal can be transformed into a number of black and white photographs of different spectral bands. The signal may be fed directly into a computer for pattern recognition such as crop distribution. The third method is to combine the signals of the wave bands into a single colour image that can be interpreted by eye. A false colour image can be obtained or be viewed through a colour-additive viewer which allows colour combinations to be varied so that when a distinctive feature such as a particular crop is identified its prominence can be increased.

This technique is used in satellite imagery in the Landsat series and gives commercial available information for all of the Earth's surface in the form of

1. Computer-compatible tapes.
2. Black-and-white photographs of four spectral bands, green, yellow to red, and two near infrared.
3. False colour composites.

This imagery is easily available and relatively inexpensive, in addition the coverage of 34 000 km^2 on a single print allows very rapid mapping of large areas. On the other hand the resolution is low and small topographic differences are not easily recognised.

Land capability classification

Most workers draw a distinction between soil and land. Whereas soil has been defined earlier, land includes the soil and a number of other attributes including vegetation, climate and water supply. This leads to concepts of land classification and land capability classification. There are two major systems of land capability classification at present in use. The first and up to the present the main system is that introduced by the USDA (Klingebiel and Montgomery 1961). The second entitled "A framework for Land Evaluation" has recently been introduced by FAO (1976). Whereas the USDA system is meant to be a general purpose system, FAO has proposed a scheme where the details are worked out for the particular crop that is to be grown and thus is specific for each case.

USDA system

This is based on the concept of limitations that affect land use. These are both permanent and temporary limitations. Land is classified mainly on the basis of permanent limitations such as angle of slope, soil depth and climate. Temporary limitations include nutrient content and poor drainage. There are three categories to the system:
1. Capability classes
2. Capability sub-classes
3. Capability units.

Capability classes

There are eight classes numbered I to VIII based on the amount of limitations that are present, with Class I having the least and Class VIII having the greatest limitations.

CLASS I The land is flat or almost flat and the soils are deep, well drained with high natural fertility or respond well to the application of fertilizers. There are few limitations to intensive use and in arid and semi-arid areas soils that can be used continuously with permanent agriculture are placed in this class.

CLASS II The soils in this class have some limitations that reduce the choice of plants or require moderate conservation practices. They cannot sustain the same intensive cropping system as in Class I or they require some conservation practice. The limiting factor may be slope, erosion hazard, shallow soil, poor structure, salinity or alkalinity or poor drainage. Conservation practices may include terracing, strip cropping, contour cultivation, special rotations.

CLASS III The soils in this class have severe limitations that reduce the choice of plants or require special conservation practices or both. Limitations result from such factors as moderately steep slope, high erosion hazard, slow permeability, shallow soil, low water-holding capacity, low fertility, moderate salinity and alkalinity, poor soil structure. Conservation practices may include terracing, strip cropping, contour cultivation, drainage, special rotations and special crops.

CLASS IV The soils in this class can be cultivated but the limiting factors are steep slopes, severe erosion hazard, shallow soils, low water-holding capacity, poor drainage, severe salinity or alkalinity. There is a severe limitations in the choice of crops.

CLASS V The soils in this class are generally not suitable for cultivation. The limiting factors other than erosion include stream overflow, short growing season, stony or rocky soils, ponded areas with poor drainage.

CLASS VI The soils have severe limitations which are the same as in Class V but they are more rigid. They are restricted in use to pasture, woodland or wildlife.

CLASS VII These soils have extreme limitations that even prevent pasture improvement.

CLASS VIII The soils in this class cannot or should not be used for crop growth. Their use is restricted to woodland, wildlife and recreation. The soils and landforms include sandy beaches, rock outcrops and steep mountain slopes.

Capability sub-classes

The four sub-classes given below are each designated by a letter:

risk of erosion	(e)	root zone limitations	(s)
wetness, drainage or overflow	(w)	climatic limitations	(c)

Thus a soil designated III(e) indicates that it is in class III because of the high risk of erosion.

Capability units

These are groups of soils that have the same potential, limitations and management responses and are indexed by Arabic numerals following the sub-class letters, for example IVw3.

This system is used in many countries or has formed the basis of many local schemes but it has a number of drawbacks including the following:

(i) it only considers arable cultivation, other forms of land use are generally ignored
(ii) individual characteristics vary according to their interaction, thus sandy soils are not always poor soils
(iii) crops require different climatic criteria
(iv) wet soils are good for rice
(v) based on negative values.

The FAO system

This system is designed to be used for specific purposes, i.e. a piece of land is classified with regard to a specific use. Therefore it is not presented as a fully developed scheme but a general one around which national or local schemes may be constructed. The scheme seems to have a great potential since it is not presented in broad generalities but is given for a specific use. The outline of the scheme is as follows:

There are four categories of decreasing generalization.

I Land suitability orders
II Land suitability classes
III Land suitability sub-classes
IV Land suitability units

I Land suitability orders

These indicate whether the land is suitable or not suitable for a specific use. There are two orders represented on maps, tables, etc. by the symbols S and N respectively.

Order S Suitable. Land on which sustained use of the kind under consideration is expected to yield benefits which justify the inputs, without unacceptable risk of damage to land resources.

Order N Not Suitable. Land which has qualities that appear to preclude sustained use of the kind under consideration.

II Land suitability classes

Any number of classes may be distinguished and indicate the extent to which the land is suitable for the defined use and are indicated by Arabic numbers. If three classes are recognised then their definitions may be as follows:

S1 Highly suitable. Land having no significant limitations to the sustained application of the defined use.

S2 Moderately suitable. Land having limitations which will reduce production levels and/or increase costs but which is physically and economically suitable for the defined use.

S3 Marginally suitable. Land having limitations which will reduce production levels and/or increase costs making it economically marginal for the defined use.

Within the Order Not Suitable there are normally two classes:

N1 Currently not suitable. Land having limitations which may be surmountable in time but which cannot be corrected with existing knowledge at presently acceptable costs and which preclude successful sustained use in the defined manner.

N2 Permanently not suitable. Land having limitations so severe as to prevent any possibility of successful sustained use in the defined manner.

III Land suitability sub-classes

Divisions of the classes based on the nature of the limitation. They are designated by lower-case letters as follows:

w = wetness limitation
t = topographic limitations
e = erosion

and written S2w, S2t, S2e. There are no sub-classes in Class S1.

IV Land suitability units

These differ in minor aspects of the management requirements. They are designated by Arabic numbers following a hyphen:

S2w–1 S2w–2

This system differ from that of the USDA in that suitability is not general purpose but is related to a specific form of land use.

Appendix 1

Examination of a soil profile

The site to be examined may be chosen subjectively so as to be fairly certain about the properties of the profile or it may be chosen objectively by statistical methods. The former method is generally used in the first stages of teaching.

After the site has been chosen the environmental factors are described and a soil pit is dug. The following are the principal environmental factors that are described and recorded on a blank form similar to that given in Fig. A1.

Location: Give the map reference or locate with reference to a fixed point.

Age: This refers to the age of the site, namely the date when the site was formed and thus the age of the soil.

Topography: This includes elevation, angle and form of slope, aspect and exposure. It also includes the type of geomorphological feature such as flood plain, pediment or river terrace.

Climate: This includes all the available data about rainfall and temperature.

Vegetation: Record the type of plant community and the frequency of the individual species.

Erosion: Examine the site carefully, and record any evidence of erosion.

Parent material: This is usually described after the pit has been dug but positive identification might be difficult and have to await the results of laboratory analyses. When the parent material is rock it can be described in standard petrological terms but when it is a sediment it can be described using their terms given on page 12 and classified using the terms given on page 13.

Drainage: Both the surface drainage and the soil drainage should be described. The surface drainage may be described as

Fig. A1 Soil Description

Date:	*Profile No:*	*Locality:* *Map reference:*	*Classification:*
Parent material *and age of site*	*Climate* *Precipitation:* *Temperature:*	*Topography* *Elevation:* *Slope:* *Aspect:* *Exposure:*	*Vegetation:* *Land use:*
Drainge *Site:* *Soil:*	*Erosion*	*Rock outcrops*	

Drawing of Profile	Depth (cm)	Horizon No. or Name	Colour	Texture	Stones	Structure and Porosity	Handling Properties	Organic Matter	Roots	Water Conditions

ponded, slow run-off, medium run-off, rapid run-off and very rapid run-off.

Digging the soil pit

This requires a certain amount of skill and patience in order to expose and reveal the true nature of the constituent horizons.

Equipment

Spade	Tape measure
Pick	Note-book or description forms
Trowel	Polythene or paper bags
Pencil	Trays or polythene sheets

Procedure

Dig a soil pit at a site chosen by oneself or by the supervisor but in either case permission must be obtained from the landowner. Generally the pit is about 2 × 1 m at the top and extends down to unaltered underlying material which usually occurs at over 1 m. Usually the pit has to be filled after the exercise is complete, therefore make two neat piles of the material from the pit. One pile contains the turfs and top soil and the other contains the subsoil, if possible place the soil on polythene or canvas sheets. Orientate the face to be examined so that it has maximum illumination, taking great care to ensure that the surface above the face to be examined is not trampled upon.

When the pit has been dug, carefully clean the face to be examined with a trowel. Now delimit the horizons and remove lumps of soil from each horizon and describe the properties given below for each horizon.

Depth and thickness:	see page 81
Name and symbol:	see page 120
Horizon boundaries:	see page 81
Colour:	see page 91
Texture:	see page 88
Rock fragments:	type, size, shape and frequency
Structure:	see page 94

Passages:	see page 104
Consistence:	see page 84
Cementation:	see page 84
Organic matter:	frequency and state of decomposition
Roots:	size and frequency
Fauna and Flora:	describe any that are visible, such as earthworms and fungal mycelium
Drainage:	see page 105
Segregations and concretions:	see page 116
Soluble salts:	see page 113
Carbonates:	see page 115
Soil reaction:	see page 108

It is possible to buy or prepare a kit to contain chemicals for testing for carbonates, pH. etc.

During the description it is essential to use a hand lens to identify a number of features including:

Fungal mycelium
Faecal material
Small crumb and granular structures
Crystalline material such as calcite and gypsum
Clay coatings
Nature of ped surfaces
Discrete pores
Small segregations and concretions

Glossary

Abrasion: The physical weathering of a rock surface by running water, glaciers or wind laden with fine particles. See Ventifact.

Absorption: The physical uptake of water and/or ions by a substance. For example, soils absorb water.

Accelerated Erosion: An increased rate of erosion caused by humans.

Accessory Minerals: Minerals occurring in small quantities in a rock whose presence or absence does not affect the true nature of the rock.

Accumulation: The build-up or increase in the amount of one or more constituents in the soil at a given position as a result of translocation. The build-up may be a residue due to the translocation of material out of the horizon or may be due to an addition of material. Usually refers to soluble substances and clay particles.

Acicular: Needle shaped.

Acidity: The hydrogen ion activity in the soil solution expressed as a pH value.

Acid Rock: An igneous rock that contains more than 60 per cent silica and free quartz.

Acid Soil: A soil with pH <6.5.

Actinomycetes: A group of organisms intermediate between the bacteria and the true fungi, mainly resembling the latter because they usually produce branched mycelium.

Adsorption: The attachment of a particle, ion or molecule to a surface. Calcium is adsorbed on to the surface of clay or humus.

Adsorption Complex: The various substances in the soil that are capable of adsorption, these are mainly the clay and humus.

Aeolian: Pertaining to or formed by wind action.

Aeolian Deposits: Fine sediments transported and deposited by wind; they include loess, dunes, desert sand and some volcanic ash.

Aeration: The process by which atmospheric air enters the soil. The rate and amount of aeration depends upon the size and continuity of the pore spaces and the degree of water logging.

Aerial Photograph: A photograph of the Earth's surface taken from an aeroplane or some other type of airborne equipment.

Aerobic: Conditions having a continuous supply of molecular oxygen.

AEROBIC ORGANISM: Organisms living or becoming active in the presence of molecular oxygen.

AGGREGATES: Discrete clusters of particles formed naturally or artificially and include such units as crumbs, granules, clods, faecal pellets, fragments of faecal pellets and concretions.

AGGREGATION: The process by which particles coalesce to form aggregates.

AGRONOMY: That part of agriculture devoted to the production of crops and soil management – the scientific utilisation of agricultural land.

ALGAE: Unicellular or multicellular plants containing chlorophyll. They are aquatic or occur in damp situations and include most seaweeds.

ALKALINE SOIL: A soil with pH >7.3.

ALLUVIAL PAN OR ALLUVIAL CONE: Sediments deposited in a characteristic fan or cone shape by a mountain stream as it flows on to a plain or flat open valley.

ALLUVIAL PLAIN: A flat area built up of alluvium.

ALLUVIAL SOIL: A general term for those soils developed on fairly recent alluvium.

ALLUVIUM: A sediment deposited by streams and varying widely in particle size. The stones and boulders when present are usually rounded or sub-rounded. Some of the most fertile soils are derived from alluvium of medium or fine texture.

AMINO ACID: An organic compound containing both the amino (NH_2) and carboxyl (COOH) groups. Amino acid molecules combine to form proteins, therefore they are a fundamental constituent of living matter. They are synthesised by autotrophic organisms, principally green plants.

AMMONIA FIXATION: Adsorption of ammonium ions by clay minerals, rendering them insoluble and non-exchangeable.

AMMONIFICATION: The production of ammonia by microorganisms through the decomposition of organic matter.

ANAEROBIC: Conditions that are free of molecular oxygen. In soils this is usually caused by excessive wetness.

ANAEROBIC ORGANISM: One that lives in an environment without molecular oxygen.

ANION: An ion having a negative charge.

ANION EXCHANGE CAPACITY: The total amount of anions that a soil can adsorb, usually expressed as meq kg^{-1} soil.

ANISOTROPIC:

1. General: possessing different physical properties in different directions
2. General: having physical properties that depend on direction
3. Minerals or parts of soils: alternately bright and dark between crossed

polars when the microscope stage is rotated. The bright position is due to the formation of interference colours. See INTERFERENCE COLOURS.

ANNELID: Red-blooded worm such as an earthworm.

ANNUAL PLANT: A plant that completes its life cycle within one year.

ARID: A term applied to a region or climate in which precipitation is too low to support crop production.

ARTHROPOD: A member of the phylum arthropoda which is the largest in the animal kingdom. It includes insects, spiders, centipedes, crabs, etc.

ASPECT: The compass direction of a slope.

AUREOLE: Halo or ring around a feature.

AUTOTROPHIC ORGANISMS: Organisms that utilise carbon dioxide as a source of carbon and obtain their energy from the sun or by oxidising inorganic substances such as sulphur, hydrogen, ammonium, and nitrate salts. The former include the higher plants and algae and the latter various bacteria, cf. HETEROTROPHIC.

AVAILABLE ELEMENTS: The elements in the soil solution that can readily be taken up by plant roots.

AVAILABLE NUTRIENTS: see AVAILABLE ELEMENTS.

AVAILABLE WATER: That part of the water in the soil that can be taken up by plant roots.

AVAILABLE WATER CAPACITY: The weight percentage of water which a soil can store in a form available to plants. It is equal to the moisture content at field capacity minus that at the wilting point.

BACTERIA: Unicellular or multicellular microscopic organisms. They occur everywhere and in very large numbers in favourable habitats such as soil and sour milk where they number many millions per gram.

BAR: 10^5 Pascal or 10^5 (Nm^{-2}).

BASALT: A fine-grained igneous rock forming lava flows or minor intrusions. It is composed of plagioclase, augite and magnetite; olivine may be present.

BASE SATURATION: The extent to which the exchange sites of a material are occupied by exchangeable basic cations; expressed as % of the cation exchange capacity.

BASIC ROCK: An igneous rock that contains less than 55 per cent silica.

BEDROCK: The solid rock at the surface of the earth or at some depth beneath the soil and superficial deposits.

BIENNIAL: A plant that completes its life cycle in two years.

BIOMASS: a) The weight of a given organism in a volume of soil that is one m^2 at the surface and extending down to the lower limit of the organism's penetration.

b) The weight of organisms in a given area or volume.

BIREFRINGENCE: The numerical difference in value between the highest and lowest refractive index of a mineral. *This is not synonymous with interference colours*. See INTERFERENCE COLOURS.

BLOCKY: Many sided with angular or rounded corners, used for describing peds.

BOG IRON ORE: A ferruginous deposit in bogs and swamps formed by oxidizing algae, bacteria or the atmosphere on iron in solution.

BOULDER CLAY: See TILL.

BS: an abbreviation for Base Saturation.

BUFFER: A substance that prevents a rapid change in pH when acids or alkalis are added to the soil, these include the clay, humus and carbonates.

BULK DENSITY: Mass per unit volume of undisturbed soil, dried to constant weight at 105°C. Usually expressed as g/cm^3.

CALCAREOUS SOIL: A soil that contains enough calcium carbonate so that it effervesces when treated with hydrochloric acid.

CALCIFICATION: Used by some to refer to the processes of calcium carbonate accumulation.

CALCITE: Crystalline calcium carbonate, $CaCO_3$. Crystallises in the hexagonal system, the main types of crystals in soils being dog-tooth, prismatic, fibrous, nodular, granular and compact.

CALICHE: A layer or horizon cemented by the deposition of calcium carbonate. It usually occurs within the soil but may be at the surface due to erosion.

CAPILLARITY: The process by which moisture moves in any direction through the fine pore spaces and as films around particles.

CAPILLARY FRINGE: The zone just above the water-table that remains practically saturated with water.

CAPILLARY MOISTURE: That amount of water that is capable of movement after the soil has drained. It is held by adhesion and surface tension as films around particles and in the finer pore spaces.

CATENA: A sequence of soils developed from similar parent material under similar climatic conditions but whose characteristics differ because of variations in relief and drainage.

CATION: An ion having a positive electrical charge.

CATION EXCHANGE: The exchange between cations in solution and cations held on the exchange sites of minerals and organic matter.

CATION EXCHANGE CAPACITY: The total potential of soils for adsorbing cations, expressed in milligram equivalents per kg of soil. Determined values depend somewhat upon the method employed.

CEC: An abbreviation of Cation Exchange Capacity.

CEMENTED: Massive and either hard or brittle depending upon the content of cementing substances such as calcium carbonate, silica, oxides of iron and aluminium, or humus.

CHALK: The term refers to either (a) soft white limestone which consists of very pure calcium carbonate and leaves little residue when treated with hydrochloric acid, sometimes consists largely of the remains of foraminifera, echinoderms, molluscs and other marine organisms, or (b) the upper or final member of the Cretaceous System.

CHAMBER: A relatively large circular or ovoid pore with smooth walls and an outlet through channels, fissures or planar pores.

CHANNEL: A tubular-shaped pore.

CHLOROSIS: The formation of pale green or yellow leaves in plants resulting from the failure of chlorophyll to develop. It is often caused by a deficiency in an essential element.

CHROMA: The relative purity of a colour directly related to the dominance of the determining wavelength. One of the three variables of colour.

CHRONOSEQUENCE: A sequence of soils that changes gradually from one to the other with time.

CIRCULARLY POLARISED LIGHT: This is produced by inserting two $\lambda/4$ mica plates between crossed polars. One plate is inserted in the 45° position in the slot of the microscope. The second plate is inserted above the polariser at right angles to the upper mica plate. The effect is that all extinction phenomena of minerals disappear and they remain bright in all positions of the stage. Only isotropic materials and basal sections appear dark.

CLAY: Either 1. Mineral material $<2\ \mu m$. 2. A class of texture. 3. Silicate clay minerals.

CLAY COATING: See COATING.

CLAY MINERAL: Crystalline or amorphous mineral material, $<2\ \mu m$ in diameter.

CLAY PAN: A middle or lower horizon containing significantly more clay than the horizon above. It is usually very dense and has a sharp upper boundary. Claypans generally impede drainage, are usually plastic and sticky when wet and hard when dry.

CLEAVAGE: The ability of a mineral or rock to split along predetermined planes.

CLIMAX VEGETATION: A fully developed plant community that is in equilibrium with its environment.

CLOD: A mass of soil produced by disturbance.

COATING: A layer of a substance completely or partly covering a surface. Coatings are composed of a variety of substances separately or in combination. They include clay coatings (clay skins), calcite coatings, whole soil coatings, etc. Coatings may become incorporated into the matrix or be fragmented.

COEFFICIENT OF LINEAR EXTENSIBILITY: The ratio of the difference between the moist and dry lengths of a clod to its dry length, (Lm-Ld)/Ld when Lm is the moist length (at $\frac{1}{3}$ atmospheres) and Ld is the air-dry length. The measure correlates with the volume change of a soil upon wetting and drying.

COLE: An abbreviation of coefficient of linear extensibility.

COLLOID: The inorganic and organic material with very fine particle size and therefore high surface area which usually exhibits exchange properties.

COLLUVIUM: Soil materials with or without rock fragments that accumulate at the base of steep slopes by gravitational action.

COMPACTION: Increase in bulk density due to mechanical forces such as tractor wheels.

COMPOSITE STRUCTURE: Any combination of different types of peds.

COMPOST: Plant and animal residues that are arranged into piles and allowed to decompose, sometimes soil or mineral fertilizers may be added.

COMPOUND STRUCTURE: Large peds such as prisms and columns that are themselves composed of smaller incomplete peds.

CONCEPT: General notion.

CONCRETION: Small hard local concentrations of material such as calcite, gypsum, iron oxide or aluminium oxide. Usually spherical or subspherical but may be irregular in shape.

CONGLOMERATE: A sedimentary rock composed mainly of rounded boulders.

CONIFEROUS FOREST: A forest consisting predominantly of cone-bearing trees with needle-shaped leaves: usually evergreen but some are deciduous, for example the larch forests (*Larix dehurica*) of central Siberia. Their greatest extent is in the wide belt across northern Canada and northern Eurasia. Coniferous forests produce soft wood which has a large number of industrial applications including paper making.

CONSISTENCE: The resistance of the soil to deformation or rupture as determined by the degree of cohesion or adhesion of the soil particles to each other.

CONSOLIDATED: A term that usually refers to compacted or cemented rocks.

CONTINUOUSLY ANAEROBIC (*very poorly drained*): A horizon that is saturated with water throughout the year, it is blue, olive or grey.

CREEP: Slow movement of masses of soil down slopes that are usually steep. The process takes place in response to gravity facilitated by saturation with water.

CROTOVINA: An animal burrow which has been filled with material from another horizon.

CROUTE CALCAIRE: A synonym for caliche.

CRUST: A surface layer of soils that becomes harder than the underlying horizon.

CUTANS: Coatings or deposits of material on the surfaces of peds, stones, etc. A common type is the clay cutan caused by translocation and deposition of clay particles on ped surfaces.

DECIDUOUS FOREST: A forest composed of trees that shed their leaves at some season of the year. In tropical areas the trees lose their leaves during the hot season in order to conserve moisture. Deciduous forests of the cool areas shed their leaves during the autumn to protect themselves against the cold and frost of winter. Deciduous forests produce valuable hardwood timber such as teak and mahogany from the tropics, oak and beech come from the cooler areas.

DEFLATION: Preferential removal of fine soil particles from the surface soil by wind. See DESERT PAVEMENT

DEFLOCCULATE: To separate or disperse particles of clay dimensions from a flocculated condition.

DELTA: A roughly triangular area at the mouth of a river composed of river transported sediment.

DENITRIFICATION: The biological reduction of nitrate to ammonia, molecular nitrogen or to the oxides of nitrogen, resulting in the loss of nitrogen into the atmosphere and therefore undesirable in agriculture.

DENUDATION: Sculpturing of the surface of the land by weathering and erosion; levelling mountains and hills to flat or gently undulating plains.

DEPOSIT: Material placed in a new position by the activity of humans or natural processes such as water, wind, ice or gravity.

DESERT CRUST: A hard surface layer in desert regions containing calcium carbonate, gypsum, or other cementing materials.

DESERT PAVEMENT: A layer of gravel or stones remaining on the surface of the ground in deserts after the removal of the fine material by wind. See DEFLATION and HAMADA.

DESERT VARNISH: A glossy sheen or coating on gravel and stones in arid regions.

DEVONIAN: A period of geological time extending from 320–280 million years BP.

DIATOMS: Algae that possess a siliceous cell wall which remains preserved after the death of the organisms. They are abundant in both fresh and salt water and in a variety of soils.

DISPERSION: The process whereby the structure or aggregation of the soil is destroyed so that each particle is separate and behaves as a unit.

DOLINE OR DOLINA: A closed depression in a karst region, often rounded or elliptical in shape, formed by the solution and subsidence of the limestone near the surface. Sometimes at the bottom there is a sink hole into which surface water flows and disappears underground.

DOMAIN: A bundle of clay particles that is only visible in crossed polarised light.

DRIFT: A generic term for superficial deposits including till (boulder-clay), outwash gravel and sand, alluvium, solifluction deposits and loess.

DRUMLIN: A small hill, composed of glacial drift with hog-back outline, oval plan, and long axis oriented in the direction of ice movement. Drumlins usually occur in groups, forming what is known as basket of eggs topography.

DRY-FARMING: A method of farming in arid and semi-arid areas without using irrigation, the land being treated so as to conserve moisture. The technique consists of cultivating a given area in alternate years allowing moisture to be stored in the fallow year. Moisture losses are reduced by producing a mulch and removing weeds. In Siberia, where melting snow provides much of the moisture for spring crops, the soil is ploughed in the autumn providing furrows in which snow can collect, preventing it from being blown away and evaporated by strong winds. Usually alternate narrow strips are cultivated in an attempt to reduce erosion in the fallow year. Dry farming methods are employed in the drier regions of India, USSR, Canada and Australia.

DUNES, SAND DUNES: Ridges or small hills of sand which have been piled up by wind action on sea coasts, in deserts and elsewhere. Barkhans are isolated dunes with characteristic crescentic forms.

ECOLOGY: The study of the interrelationships between individual organisms and between organisms and their environment.

ECOSYSTEM: A group of organisms interacting among themselves and with their environment.

EDAPHIC: (1) Of or pertaining to the soil. (2) Influenced by soil factors.

EDAPHOLOGY: The study of the relationships between soil and organisms including the use of the land by humans.

EFFLORESCENCE: The accumulation of dissolved substance (usually simple salts) at a surface due to evaporation.

ELUVIAL HORIZON: A horizon from which material has been removed either in solution or suspension.

ELUVIATION: Removal of material from the upper horizon in solution or suspension.

EQUATORIAL FOREST OR TROPICAL RAIN FOREST: A dense, luxuriant, evergreen forest of hot, wet, equatorial regions containing many trees of tremendous heights, largely covered with lianas and epiphytes. Individual species of trees are infrequent but they include such valuable tropical hardwoods as mahogany, ebony and rubber. Typical equatorial forests occur in the Zaire and Amazon basins and southeastern Asia.

EROSION: The removal of material from the surface of the land by weathering, running water, moving ice, wind and mass movement.

EROSION PAVEMENT: A layer of gravel or stones left on the surface of the ground after the removal of the fine particles by erosion.

ESKER: A long narrow ridge, chiefly of gravel and sand, formed by a melting glacier or ice sheet.

EUTROPHIC: Containing an optimum concentration of plant nutrients.

EVAPOTRANSPIRATION: The combined processes of evaporation and transpiration.

EXCESSIVELY AEROBIC: A horizon which is usually too dry to support adequate plant growth.

EXCESSIVELY DRAINED: A soil that loses water very rapidly because of rapid percolation.

EXCHANGEABLE CATION: A cation such as calcium that is adsorbed onto a surface, usually clay or humus and is capable of being easily replaced by another cation such as potassium. Exchangeable cations are readily available to plants.

EXFOLIATION: A weathering process during which thin layers of rock peel off from the surface. This is caused by heating of the rock surface during the day and cooling at night leading to alternate expansion and contraction. This process is sometimes termed "onion skin weathering".

EXTINCTION: The position at which a crystal goes black in crossed polarised light.

EXTINCTION ANGLE: The angle at which a crystal goes black in crossed polarised light.

FABRIC: See SOIL FABRIC.

FAECAL MATERIAL: The various types of faeces or excrement produced by soil fauna.

FAECAL PELLETS: Rounded and subrounded aggregates of faecal material produced by the soil fauna.

FALLOW: Leaving the land uncropped for a period of time. This may be to accumulate moisture, improve structure or induce mineralisation of nutrients.

FAMILY: One of the categories in soil classification intermediate between the great soil group and the soil series.

Fen Peat: Peat that is neutral to alkaline due to the presence of calcium carbonate.

Ferralitisation: Used by some to refer to the processes of formation of ferralitic soils. This term is not specific and should not be used.

Fertilizer: A material that is added to the soil to supply one or more plant nutrients in a readily available form.

Field Capacity or Field Moisture Capacity: The total amount of water remaining in a freely drained soil after the excess has flowed into the underlying unsaturated soil. It is expressed as a percentage of the oven-dry soil.

Fine Material: Soil material in thin sections composed of particles less than 2 μm which are difficult or impossible to resolve with the petrological microscope.

Fine Texture: Containing >35 per cent clay.

Flood Plain: The land adjacent to a stream, built of alluvium and subject to repeated flooding.

Fluvio-Glacial: See Glacio-Fluvial Deposits.

Fragment: A small mass of soil produced by disturbance.

Freely Drained: A soil that allows water to percolate freely.

Friable: A term applied to soils that when either wet or dry crumble easily between the fingers.

Fulvic Acid: The mixture of organic substances remaining in solution upon acidification of a dilute alkali extract of soil.

Fungi: Simple plants that lack chlorophyll and composed of cellular filamentous growth known as hyphae. Many fungi are microscopic but their fruiting bodies, viz. mushrooms and puffballs are quite large.

Gastropod: A member of the Gastropoda class of molluscs which includes snails and slugs.

Geomorphology: The study of the origin of physical features of the Earth, as they are related to geological structure and denudation.

Gilgai: A distinctive microrelief of knolls and basins that develops on clay soils that exhibit a considerable amount of expansion and contraction in response to wetting and drying.

Glacial Drift: Material transported by glaciers and deposited directly from the ice or from the melt water.

Glacier: A large mass of ice that moves slowly over the surface of the ground or down a valley. They originate in snow fields and terminate at lower elevations in a warmer environment where they melt.

Glacio-Fluvial Deposits: Material deposited by meltwaters coming from a glacier. These deposits are variously stratified and may form outwash plains, deltas, kames, eskers, and kame terraces. See Glacial Drift and Till.

GLEISATION: See GLEYING.

GLEYING: The reduction of iron in an anaerobic environment leading to the formation of grey or blue colours.

GRANITE: An igneous rock that contains quartz, feldspars and varying amounts of biotite and muscovite.

GRAVITATIONAL WATER: The water that flows freely through soils in response to gravity.

GREAT SOIL GROUP: One of the categories in soil classification.

GROUNDWATER-TABLE: The upper limit of the ground water.

GULLY: A shallow steep-sided valley that may occur naturally or be formed by accelerated erosion.

GULLY EROSION: A form of catastrophic erosion that forms gullies.

GYTTJA: Peat consisting of faecal material, strongly decomposed plant remains, shells of diatoms, phytoliths, and fine mineral particles. Usually forms in standing water.

HALOMORPHIC SOIL: A soil containing a significant proportion of soluble salts.

HALOPHYTE: A plant capable of growing in salty soil; i.e. a salt tolerant plant.

HALOPHYTIC VEGETATION: Vegetation that requires or tolerates saline conditions.

HAMADA: An accumulation of stones at the surface of deserts, formed by the washing or blowing away of the finer material.

HARDPAN: A horizon cemented with organic matter, silica, sesquioxides, or calcium carbonate. Hardness or rigidity is maintained when wet or dry and samples do not slake in water.

HEAVY SOIL (Obsolete): A soil that has a high content of clay and is difficult to cultivate.

HETEROTROPHIC ORGANISMS: Those that derive their energy by decomposing organic compounds, cf. AUTOTROPHIC

HOLOCENE PERIOD: The period extending from 10,000–0 years BP.

HORIZON: Relatively uniform material that extends laterally, continuously or discontinuously throughout the pedounit; runs approximately parallel to the surface of the ground and differs from the related horizons in many chemical, physical and biological properties.

HUE: The dominant spectral colour and one of the three colour variables.

HUMIC ACID: Usually refers to the mixture of ill-defined dark organic substances precipitated upon acidification of a dilute alkali extract of soil. Some workers use it to include only the alcohol-insoluble portion of the precipitate.

HUMIFICATION: The decomposition of organic matter leading to the formation of humus.

HUMIN: Usually applied to that part of the organic matter that remains after extraction with dilute alkali.

HUMUS: The well-decomposed, relatively stable part of the organic matter found in aerobic soils.

HYDRATION: The process whereby a substance takes up water.

HYDRAULIC CONDUCTIVITY: The rate at which water will move through soil in response to a given potential gradient.

HYDROLOGIC CYCLE: Disposal of precipitation from the time it reaches the soil surface until it re-enters the atmosphere by evapotranspiration to serve again as a source of precipitation.

HYDROLYSIS: In soils it is the process whereby hydrogen ions are exchanged for cations such as sodium, potassium, calcium and magnesium.

HYDROMORPHIC SOIL: Soils developed in the presence of excess water.

HYGROSCOPIC WATER: Water that is adsorbed on to a surface from the atmosphere.

IGNEOUS ROCK: A rock formed by the cooling of molten magma including basalt and granite.

ILLUVIAL HORIZON: A horizon that receives material in solution or suspension from some other part of the soil.

ILLUVIATION: The process of movement of material from one horizon and its deposition in another horizon of the same soil; usually from an upper horizon to a middle or lower horizon in the pedounit. Movement can also take place laterally.

IMMATURE SOIL: Lacking a well-developed pedo-unit.

IMPEDED DRAINAGE: Restriction of the downward movement of water by gravity.

IMPERFECTLY DRAINED: A soil that shows a small amount of reduction of iron due to short periods of waterlogging.

IMPERVIOUS: Not easily penetrated by roots or water.

INCOMPLETE STRUCTURE: Aggregates joined to each other by narrow necks. (FitzPatrick, 1980).

INFILTRATION: The process whereby water enters the soil through the surface.

INSELBERG: (pl. Inselberge) A steep-sided hill composed predominantly of hard rock and rising abruptly above a plain; found mainly in tropical and subtropical areas.

INTERFERENCE COLOURS: The colours of the Newton scale that are formed when a birefringent mineral or some plant material is examined between crossed polarizers. See BIREFRINGENCE.

INTERGLACIAL PERIOD: A relatively mild period occurring between two glacial periods.

INTERGRADE: A soil which contains the properties of two distinctive and genetically different soils.

INTERSTADIAL PERIOD: A slightly warmer phase during a glacial period.

INTRAZONAL SOILS: One of the three orders of the zonal system of soil classification. They have well developed characteristics resulting from the dominant influence of a local factor such as topography and parent material.

ISOMORPHOUS REPLACEMENT: The replacement of one ion by another in the crystal lattice without changing the structure of the mineral.

ISOTROPIC: Not visible in crossed polarised light. See ANISOTROPIC.

KARST TOPOGRAPHY: An irregular land surface in a limestone region. The principal features are depression (e.g. dolines which sometimes contain thick soils which have been washed off the rest of the surfaces leaving them bare and rocky). Drainage is usually by underground streams.

KROTOVINA: see CROTOVINA.

LACUSTRINE: Pertaining to lakes.

LACUSTRINE DEPOSIT: Materials deposited by lake waters.

LANDSLIDE OR LANDSLIP: The movement down the slope of a large mass of soil or rocks from a mountain or cliff. Often occurs after torrential rain which soaks into the soil making it heavier and more mobile. Earthquakes and the undermining action of the sea are also causative agents.

LATERISATION: Used by some to refer to the processes of formation of laterite or red and yellow tropical soils. This term is not specific and should not be used.

LATTICE STRUCTURE: The orderly arrangement of atoms in crystalline material.

LEACHING: The washing out of material from soil, both in solution and suspension.

LIGHT SOIL: (obsolete) A soil which has a coarse texture and is easily cultivated.

LIME: Compounds of calcium used to correct the acidity in soils.

LITTER: The freshly fallen plant material occurring on the surface of the ground.

LODGING: The collapse of top-heavy plants, particularly grain crops because of excessive growth or beating by rain.

LOESS: An aeolian deposit composed mainly of silt which originated in arid regions, from glacial outwash or from alluvium. It is usually of yellowish brown colour and has a widely varying calcium carbonate content. In the USSR, loess is regarded as having been deposited by water.

LYSIMETER: Apparatus installed in the soil for measuring percolation and leaching.

MACROELEMENT: Elements such as nitrogen that are needed in large amounts for plant growth.

MACRONUTRIENT: See MACROELEMENT.

MACROPORES: Pores $>100\ \mu m$ in diameter.

MANGROVE SWAMP: A dense jungle of mangrove trees which have the special adaptation of extending from their branches long arching roots which act as anchors and form an almost impenetrable tangle. They occur in tropical and subtropical areas, particularly near to the mouths of rivers.

MANURE: Animal excreta with or without a mixture of bedding or litter.

MATRIX: The fine material (generally $<2\ \mu m$) forming a continuous phase and enclosing coarser material and/or pores.

MATURE SOIL: A well developed soil usually with clearly defined horizonation.

MERISTEM: The region of active cell-division in plants, it is the tips of stems and roots in most plants. The cells so formed then become modified to form the various tissues such as the epidermis and cortex.

MESOFAUNA: Small organisms such as worms and insects.

METAMORPHIC ROCK: A rock that has been derived from other rocks by heat and pressure. The original rock may have been igneous, sedimentary or another metamorphic rock.

MICROCLIMATE: The climate of a very small region.

MICROELEMENT: Those elements that are essential for plant growth but are required only in very small amounts.

MICROFAUNA: The small animals that can only be seen with a microscope; they include protozoa, nematodes, etc.

MICROFLORA: The small plants that can only be seen with a microscopic; they include algae, fungi, bacteria, etc.

MICRONUTRIENT: See MICROELEMENT.

MICROPORES: Pores 5–30 μm in diameter.

MICROORGANISM: The members of the microflora and microfauna that can only be seen with a microscope.

MICRORELIEF: Small differences in relief that have differences in elevation up to about 2 m.

MILLIEQUIVALENT: A thousandth of an equivalent weight.

MINERALISATION: The change of an element in an organic form to an inorganic form by microorganisms.

MINERAL SOIL: A soil that is composed predominantly of mineral material cf. ORGANIC SOIL.
MITES: Very small members of the arachnid which includes spiders; they occur in large numbers in many organic surface soils.
MODER: A kind of decomposition and humus formation which produces advanced but incomplete humification of the remains of organism due to good aeration.
MOR: An accumulation of acid organic matter at the soil surface beneath forest.
MORAINE: Any type of constructional topographic form consisting of till and resulting from glacial deposition.
MOTTLING: Patches or spots of different colours usually used for the colour pattern developed due to partial anaerobism.
MULCH: A loose surface horizon that forms naturally or may be produced by cultivation and consists of either inorganic or organic materials.
MULL: A crumbly intimate mixture of organic and mineral material formed mainly by worms, particularly earthworms.

NEUTRAL SOIL: A soil with pH values 6.5–7.3.
NITRIFICATION: The oxidation of ammonia to nitrite and nitrite to nitrate by microorganisms.
NITROGEN FIXATION: The transformation of elemental nitrogen to an organic form by microorganisms.
NON-SILICATE: Rock forming minerals that do not contain silicon.

ONION SKIN WEATHERING: See EXFOLIATION.
ORGANIC SOIL: A soil that is composed predominantly of organic matter, usually refers to peat.

PANS: Soil horizons that are strongly compacted, cemented or have a high content of clay.
PARENT MATERIAL: The original state of the soil. The relatively unaltered lower material in soils is often similar to the material in which the horizons above have formed.
PEAT: An accumulation of dead plant material often forming a layer many metres deep. It is only slightly decomposed due to being completely waterlogged.
PED: A single individual naturally occurring soil aggregate such as a granule or prism cf. CLOD or FRAGMENT.
PEDOGENESIS: The natural process of soil formation.
PEDOLOGY: The study of soils as naturally occurring phenomena taking into account their composition distribution and method of formation.

Pedounit: A selected column of soil containing sufficient material in each horizon for adequate laboratory characterisation.

Pedoturbation: All mixing of soil components that is not caused by illuviation.

Peneplain: A large flat or gently undulating area. Its formation is attributed to progressive erosion by rivers and rain, which continues until almost all the elevated portions of the land surface are worn down. When a peneplain is elevated, it may become a plateau which then forms the initial stages in the development of a second peneplain.

Perched Water-Table: The upper limit of perched water. See Perched impermeable layer such as a pan or a high content of clay.

Perched Water-table: The upper limit of perched water. See Perched Water.

Percolation: (soil water) The downward or lateral movement of water through soil.

Perennial: A plant that continues to grow from year to year.

Permafrost: Permanently frozen subsoil.

Permanent Wilting Point: See Wilting Point.

Permeability: The ease with which air, or plant roots penetrate into or pass through a specific horizon.

pH: The negative logarithm of the hydrogen ion concentration of a solution. It is the quantitative expression of the acidity and alkalinity of a solution and has a scale that ranges from about 0 to 14. pH 7 is neutral, <7 is acid and >7 is alkaline.

pH, Soil: The negative logarithm of the hydrogen ion concentration of a soil solution. The degree of acidity (or alkalinity) of a soil expressed in terms of the pH scale, from 2 to 10.

Physical Weathering: The comminution of rocks into smaller fragments by physical forces such as frost action and exfoliation.

Physiological Drought: A temporary daytime state of drought in plants due to the losses of water by transpiration being more rapid than uptake by roots even although the soil may have an adequate supply. Such plants usually recover during the night.

Phytolith: Opaline formation in plant tissue that remains in the soil after the softer plant tissue has decomposed.

Plagioclimax: A plant community which is maintained by continuous human activity of a specific nature such as burning or grazing.

Plastic: A moist or wet soil that can be moulded without rupture.

Platy: Soil aggregates that are horizontally elongated.

Pleistocene Period: The period following the Pliocene period, extending from 2,000,000–10,000 years BP. In Europe and North America, there is evidence of four or five periods of intense cold during this period, when large areas of the land surface were covered by ice – glacial periods. During the interglacial periods the climate ameliorated and the glaciers retreated.

PLEOCHROISM: (minerals). The changes in colour when some transparent minerals are rotated in plane polarized light. It is expressed in terms of the nature and intensity of the colour change.

PLUVIAL PERIOD: A period of hundreds of thousands of years of heavy rainfall.

PODZOLISATION: Used by some to refer to the process of formation of a Podzol. This term is not specific and should not be used.

POLDER: A term used in Holland for an area of land reclaimed from the sea or a lake. A dyke is constructed around the area which is then drained by pumping the water out. Polders form valuable agricultural land or pasture land for cattle.

POLYGENIC SOIL: A soil that has been formed by two or more different and contrasting processes so that all the horizons are not genetically related.

POORLY DRAINED: See STRONGLY ANAEROBIC.

PORE: A discrete volume of soil atmosphere completely surrounded by soil (cf. PORE SPACE).

PORE SPACE: The continuous and interconnecting spaces in soils.

POROSITY: The volume of the soil mass occupied by pores and pore space.

PRIMARY MINERAL: 1. A mineral such as feldspar or a mica which occurs or occurred originally in an igneous rock.
2. Any mineral which occurs in the parent material of the soil.

PROFILE: A vertical section through a soil from the surface into the relatively unaltered material.

PSEUDOMORPH: A mineral having the characteristic outward form of another mineral or object it replaces.

PUDDLE: To destroy the structure of the surface soil by physical methods such as the impact of rain drops, poor cultivation with implements and trampling by animals.

QUATERNARY ERA: The period of geological time following the Tertiary Era, it includes the Pleistocene and Holocene periods and extends from 2,000,000–0 years BP.

RAINFALL INTERCEPTION: The interception and accumulation of rainfall by the foliage and branches of vegetation.

RAIN SPLASH: The redistribution of soil particles on the surface by the impact of rain drops. On slopes this can cause a large amount of erosion.

RAIN SPLASH EROSION: See RAIN SPLASH.

RAISED BEACH: A beach raised by earth movement thus forming a narrow coastal plain. There may be raised beaches at different levels resulting from repeated earth movement.

RAW HUMUS: A humus form consisting predominantly of well preserved, though often fragmented plant remains with few faecal pellets.

RHIZOSPHERE: The soil close to plant roots where there is usually an abundant and specific microbiological population.

RILL: A small intermittent water course with steep sides.

RILL EROSION: The formation of rills as a consequence of poor cultivation.

REGOLITH: The unconsolidated mantle of weathered rock, soil and superficial deposits overlying solid rock.

RUBIFACTION: The development of a red colour in soil – reddening.

SALINE SOIL: A soil containing enough soluble salts to reduce its fertility.

SALINISATION: The process of accumulation of salts in soil.

SAND: Mineral or rock fragments that range in diameter from 2–0.02 mm in the international system or 2–0.05 mm in the USDA system.

SATURATED FLOW: The movement of water in a soil that is completely filled with water.

SCLEROTIA: Spherical resting stages of fungi.

SECONDARY MINERAL: Those minerals that form from the material released by weathering. The main secondary minerals are the clays and oxides.

SEDIMENTARY ROCK: A rock composed of sediments with varying degrees of consolidation. The main sedimentary rocks include sandstones, shales, conglomerates and some limestones.

SELF-MULCHING SOIL: A soil with a naturally formed well aggregated surface which does not crust and seal under the impact of raindrops.

SESQUIOXIDES: Usually refers to the combined amorphous oxides of iron and aluminium.

SHEET EROSION: The gradual and uniform removal of the surface soil by water without forming any rills or gullies.

SILICATES: Rock forming minerals that contain silicon.

SILT: Mineral particles that range in diameter from 0.02–0.002 mm in the international system or 0.05–0.002 mm in the USDA system.

SLICKENSIDE: The polished surface that forms when two peds rub against each other when some soils expand in response to wetting.

SLICKSPOT: Small areas of surface soil that are slick when wet because of alkalinity or high exchangeable sodium.

SOIL: The natural space-time continuum occurring at the surface of the Earth and supporting plant life.

SOIL ANISOTROPY: The occurrence of a vertical horizon sequence in soils causes vertical anisotropy to be an essential characteristic. Frequently this vertical anisotropy can also be observed in thin sections. See ANISOTROPIC.

SOIL AUGER: A tool used for boring into the soil and withdrawing small samples for field or laboratory examination.

SOIL ERRATICS:

1. Fragments of horizons or other soil features transported and incorporated in superficial deposits in which a soil may have formed or is forming.
2. Part of a previously existing horizon preserved within a subsequently formed horizon.

SOIL FABRIC: The arrangement, size, shape and frequency of the individual soil constituents excluding pores.

SOIL HORIZON: See HORIZON.

SOIL MONOLITH: A vertical section through the soil preserved with resin and mounted for display.

SOIL PROFILE: A section of two dimensions extending vertically from the Earth's surface so as to expose all the soil horizons and a part of the relatively unaltered underlying material.

SOIL SURVEY: The systematic examination and mapping of soil.

SOLIFLUCTION: Slow flow of material on sloping ground, characteristic of, though not confined to, regions subjected to alternate periods of freezing and thawing.

SOLUM: The part of the soil above the relatively unaltered material.

SPHERICITY: relates to the overall shape of a feature irrespective of the sharpness of its edges and is a measure of the degree of its conformity to a sphere.

SPRINGTAILS: Very small insects that live in the surface soil.

STRIP CROPPING: The practice of growing crops in strips along the contour in an attempt to reduce run off, thereby preventing erosion or conserving moisture.

STRONGLY ANAEROBIC: (Poorly drained) soil that remains very wet or waterlogged for long periods of the year and as a result develops a mottled pattern of greys and browns.

STRUCTURE: The spatial distribution and total organisation of the soil system as expressed by the degree and type of aggregation and the nature and distribution of pores and pore space.

SUBHEDRAL: Minerals with partly developed crystallographic form.

SYMBIOSIS: Two organisms that live together for their mutual benefit. Fungus and alga that forms a lichen or nitrogen fixing bacteria living in roots are examples of symbiosis. The individual organisms are called symbionts.

TALUS: Angular rock fragments that accumulate by gravity at the foot of steep slopes of cliffs.

TECTONIC: Rock structures produced by movements in the Earth's crust.

TERRACE: A broad surface running along the contour. It can be a natural phenomenon or specially constructed to intercept run off – thereby preventing erosion and conserving moisture. Sometimes they are built to provide adequate rooting depth for plants.

TERTIARY PERIOD: The period of time extending from 75,000,000–2,000,000 years BP.

THERMOPHILIC BACTERIA: Bacteria which have optimum activity between about 45° and 55°C.

THORN FOREST: A deciduous forest of small, thorny trees, developed in a tropical semi-arid climate.

TILE DRAIN: Short lengths of concrete or pottery pipes placed end to end at a suitable depth and spacing in the soil to collect water from the soil and lead it to an outlet.

TILL: An unstratified or crudely stratified glacial deposit consisting of a stiff matrix of fine rock fragments and old soil containing subangular stones of various sizes and composition, many of which may be striated (scratched). It forms a mantle from less than 1 m to over 100 m in thickness covering areas which carried an ice-sheet or glaciers during the Pleistocene and Holocene periods.

TILL PLAIN: A level or undulating land surface covered by glacial till.

TILTH: The physical state of the soil that determines its suitability for plant growth taking into account texture, structure, consistence and pore space. It is a subjective estimation and is judged by experience.

TOPOSEQUENCE: A sequence of soils whose properties are determined by their particular topographic situation.

TOXIC SUBSTANCE: A substance that is present in the soil or the above ground atmosphere that inhibits the growth of plants and ultimately may cause their death.

TRANSLOCATION: Migration of material in solution or suspension from one horizon to another.

TRIASSIC: A period of geological time extending from 190,000,000–150,000,000 years BP.

TROPICAL RAIN FOREST: See EQUATORIAL FOREST.

UNAVAILABLE NUTRIENTS: Plant nutrients that are present in the soil but cannot be taken up by the roots because they have not been released from the rock or minerals by weathering or from organic matter by decomposition.

UNAVAILABLE WATER: Water that is present in the soil but cannot be taken up by plant roots because it is strongly adsorbed onto the surface of particles.

UNCONSOLIDATED: Sediments that are loose and not hardened.

ULTRAMICROPORES: Pores $< 5\ \mu m$ in diameter.

UNSATURATED FLOW: The movement of water in a soil that is not completely filled with water.

VALUE: The relative lightness or intensity of colour, one of the three colour variables.

VARNISH (DESERT): A dark shiny coating on stones in deserts, probably composed of compounds of iron and manganese (cf. DESERT VARNISH).

VARVE: A layer representing the annual deposit of a sediment, it usually consists of a lighter and darker portion due to the change in rate of decomposition during the year. The material may be of any origin but the term is most often used in connection with glacial lake sediments.

VENTIFACT: A pebble facetted or moulded by wind action, usually forms in polar and desert areas. The flat facets meet at sharp angles.

VERY POORLY DRAINED: A soil that remains wet and waterlogged for most of the year so that most of the horizons are blue, olive or grey due to the reducing conditions.

VOLCANIC ASH (Volcanic Dust): Fine particles of lava ejected during a volcanic eruption. Sometimes the particles are shot high into the atmosphere and carried long distances by the wind.

WATERLOGGED: Saturated with water.

WATER-TABLE (Ground): The upper limit in the soil or underlying material permanently saturated with water.

WATER-TABLE PERCHED: See PERCHED WATER-TABLE.

WEAKLY ANAEROBIC: A horizon that is anaerobic for short periods and moist for long periods. The colours are less bright than aerobic horizons and they are usually marbled or weakly mottled.

WEATHERING: All the physical, chemical and biological processes that cause the disintegration of rocks at or near the surface.

WELL DRAINED: See AEROBIC.

WILTING POINT: The percentage by weight of water remaining in the soil when the plants wilt permanently.

XEROPHYTES: Plants that grow in extremely dry areas.

References

ALEXANDER, M., 1977. *Introduction to Soil Microbiology*. John Wiley, p. 467.

BIBBY, J. S. and D. MACKNEY, 1969. *Land use capability classification*, Tech. Mono. No. 1 Soil Survey of Great Britain, p. 27.

BRADY, N. C., 1984. *The Nature and Properties of Soils* (9th edn). Macmillan: New York; Collier Macmillan: London, p. 750.

BRIDGES, E. M., 1978. *World Soils* (2nd edn). Cambridge University Press, p. 128.

FAO., 1976. *A Framework for Land Evaluation*, Soils Bulletin No. 32 FAO, Rome, p. 72.

FAO-UNESCO., *Soil Map of the World*, Vol. 1. Legend, Paris, p. 59.

FITZPATRICK, E. A., 1980. *Soils, Their Formation, Classification and Distribution*. Longman: London and New York, p. 353.

FITZPATRICK, E. A., 1984. *Micromorphology of Soils*. Chapman and Hall: London, New York, p. 433.

GREENWOOD, D. J., 1970. 'Soil aeration and plant growth', *Prog. App. Chem.* **55**, 423–31.

HUXLEY, J. G., 1943. *TVA Adventure in Planning*. Architectural Press: London, p. 142.

JENNY, H., 1941. *Factors of Soil Formation*. McGraw-Hill, p. 142.

KLINGEBIEL, A. A. and P. H. MONTGOMERY, 1961. *Land Capability Classification*. USDA Agricultural Handbook No. 210, p. 21.

PIZER, N. H., 1961. *The Practical Application of Knowledge of Soils*. Ministry of Agriculture, Fisheries and Food, Agricultural Land Services, Technical Report No. 8, p 15–20.

RUSSELL, Sir E. J., 1957. *The World of the Soil*. (New Naturalist Series). Collins: London, p. 237.

SIMPSON, K., 1983. *Soil*. Longman: London and New York, p. 238.

USDA, 1971. *Guide for Interpreting Engineering Uses of Soils*. p. 87.

USDA, 1975. *Soil Taxonomy*. Agricultural Handbook No. 436, p. 754.

WALLWORK, J. A., 1970. *Ecology of Soil Animals*. McGraw-Hill: London, p. 283.

WISCHMEIER, W. H. and D. D. SMITH, 1978. *Predicting Rainfall Erosion Losses – a guide to conservation planning*. USDA Agricultural Handbook 537, Washington, D.C., USA Government Printing Office, p. 58.

YAALON, D. H., 1960. 'Some implications of fundamental concepts of pedology in soil classification' *Trans. 7th Inter. Cong. Soil Sci.*, IV, pp. 119–23.

See also *Memoirs of Soil Survey of England and Wales*. Harpenden. *Memoirs of Soil Survey of Scotland*. Macaulay Institute, Aberdeen.

Index